Ecology and Behavior of Gulls

EDITORS

JUDITH LATTA HAND

17615 Parlange Place
San Diego, CA 92128

WILLIAM E. SOUTHERN

Dept. of Biological Sciences
Northern Illinois University
DeKalb, IL 60115

and

KEES VERMEER

Canadian Wildlife Service
P.O. Box 6000
Sydney, British Columbia V8L 4B2

Proceedings of an International Symposium of the
COLONIAL WATERBIRD GROUP
and the
PACIFIC SEABIRD GROUP
San Francisco, California
6 December 1985

Studies in Avian Biology No. 10

A PUBLICATION OF THE COOPER ORNITHOLOGICAL SOCIETY

Cover Photograph: California Gulls (*Larus californicus*) on breeding grounds at Mono Lake, California, by Joseph R. Jehl, Jr.

STUDIES IN AVIAN BIOLOGY

Edited by

FRANK A. PITELKA

at the

Museum of Vertebrate Zoology
University of California
Berkeley, CA 94720

Studies in Avian Biology is a series of works too long for *The Condor,* published at irregular intervals by the Cooper Ornithological Society. Manuscripts for consideration should be submitted to the editor-designate, Joseph R. Jehl, Jr., Sea World Research Institute, 1700 South Shores Road, San Diego, CA 92109. Style and format should follow those of previous issues.

Price: $18.50 including postage and handling. All orders cash in advance; make checks payable to Cooper Ornithological Society. Send orders to James R. Northern, Assistant Treasurer, Cooper Ornithological Society, Department of Biology, University of California, Los Angeles, CA 90024.

ISBN: 0-935868-31-3
Library of Congress Catalog Card Number 87-71187
Printed at Allen Press, Inc., Lawrence, Kansas 66044
Issued 9 June 1987

CONTENTS

* Abstracts only.

PREFACE

At the 1985 joint meeting of the Pacific Seabird Group and the Colonical Waterbird Group held 4–8 December at the Financial District Holiday Inn in San Francisco, one day was devoted to a symposium on gull biology. Speakers represented a broad spectrum of interests in current studies of gull ecology and behavior, coming from Canada, Mexico, Iceland, The Netherlands, and various parts of the United States. Altogether, 23 papers were delivered, and 21 appear in the text which follows, 11 as full papers and 10 as abstracts. That eleven other papers on gulls were also presented at the 4-day meeting reflects the continuing broad appeal of gulls and the places gulls inhabit for studies of a wide variety of ecological, behavioral and evolutionary problems.

It was for this reason that, as series editor, I urged the three organizers of this symposium to include in its published form an introductory paper scanning the current horizon of recent work. The result is Bill Southern's introductory overview. The guest editors and I hope that this review along with the symposium papers will provide useful background and cues for new work.

At an editorial staff meeting I attended for another publication, a fairly strong view was expressed questioning the utility of any "grab-bag" collection of papers on birds of a particular major taxon. I disagreed, also fairly strongly. General principles and rules cutting across species rest on in-depth studies of patterns in individual major taxa. The diversity of birds, their popularity as subjects for field studies, and their importance in the explosion of literature in ecology, behavior and evolution create a need for periodic assessment of directions and goals that taxon-oriented symposia can provide. Of course, the success of such symposia varies, but the need remains.

More particularly, a major group, such as gulls, displays a characteristic set of shared morphologic, physiologic, behavioral, and life-history features differing fundamentally from other such sets in the class Aves. These provide critical bases for between-population and between-species comparisons useful in the analysis of factors governing a group's success in functional, demographic, and evolutionary terms. For other major taxa, less well known and differing in basic features of design, symposia such as the present one do, or should, provide useful perspective for both choice and focus of research and for the testing of theory. These are among the worthy goals that symposia on the biology of major taxa can serve, and I believe this one does its share.

This is the third PSG symposium devoted to marine birds published in Studies in Avian Biology. Earlier ones dealt with tropical seabirds (1983, SAB 8) and shorebirds (1979, SAB 2). At least one new one is in the planning stage.

Frank A. Pitelka

11 February 1987

ACKNOWLEDGMENTS

The editors express special appreciation for the considerable and critical assistance given by the following reviewers of manuscripts for this symposium. Names of those reviewing more than one manuscript are followed by an asterisk.

D. W. Anderson
D. A. Bell
H. Blokpoel
B. M. Braune
J. Burger
R. G. Butler*
R. W. Butler
J. W. Chardine
M. Conover
M. Coulter
R. M. Evans
M. Fitch
G. A. Fox
D. M. Fry
M. Gochfeld
L. A. Hanners
A. B. Harper
W. Hoffman
T. R. Howell
G. L. Hunt, Jr.
D. B. Irons
J. R. Jehl, Jr.*
P. A. Jones
S. A. Mahoney
M. K. McNicholl
R. D. Morris*
W. A. Montevecchi
E. C. Murphy*
I. C. T. Nisbet
S. R. Patton
R. Pierotti*
J. B. Ryder
G. A. Sanger
R. W. Schreiber
G. W. Shugart
L. K. Southern
A. L. Spaans
J. A. Spendelow
B. Termaat
N. A. M. Verbeek
D. W. Winkler*
K. Winnett-Murray

Studies in Avian Biology No. 10:1–7, 1987.

GULL RESEARCH IN THE 1980s: SYMPOSIUM OVERVIEW

WILLIAM E. SOUTHERN[1]

Symposia are now a regular feature of the annual meetings of scientific organizations. Two approaches are available for selecting themes for symposia. The subject may be a concept, such as the mechanisms of bird orientation, and researchers working on an array of species present results tied together by the unifying conceptual thread. The other option is to use a taxon as the common denominator and have the investigators discuss a variety of concepts as they apply to one or more closely related species. Both types of symposia have their advantages. The organizers of the 1985 First Joint Meeting of the Pacific Seabird Group (PSG) and the Colonial Waterbird Group (CWG) considered it an appropriate time to review the status of gull research in the 1980s. Presented herein are 11 papers and 10 abstracts reflecting current larid research and the approaches investigators are taking.

Gulls stand out as appropriate subjects for consideration at a scientific meeting because of their relationship with man in the past, present and, more than likely, in the future. During the nineteenth century egging activities, the feather trade, reduction in fish populations, use of nearshore islands for livestock grazing and other human endeavors lowered gull populations in North America (Graham 1975). Protection in the form of state, national and international legislation early in the twentieth century resulted in gradual population increases until about midway through this century. Then there was a continent-wide explosion in the population of several gull species (e.g., Drury and Kadlec 1974, Ludwig 1974). Environmental changes that probably made these population changes possible included the introduction of forage fishes (smelt and alewives) in the Great Lakes, the operation of large landfills throughout the winter ranges of North American gulls, construction of dredge-spoil islands, and the construction of new resting habitat (numerous ponds and reservoirs) throughout the ranges of some species.

In the 1980s, gull populations have become large or concentrated enough to result in conflicts with man. The increase in competition between gulls and man has added a practical component to gull research. In order to develop management strategies that are resource sensitive while also providing for man's environmental requirements, we must possess an in-depth understanding of the species involved, including their breeding biology, habitat requirements, food habits, and long-term responses to environmental change. The papers presented in this symposium contribute significantly to the development of a data base that is essential for resource managers. In addition, many of the papers address more theoretical aspects of behavioral ecology for which gulls are ideal subjects because of their colonial nesting habits and their tendency to use nest sites accessible to investigators.

Gulls as a group also have served as the subjects of basic research that has contributed to the formulation of many major biological concepts. Such studies have expanded our understanding of motivational systems (Tinbergen 1953, Baerends and Drent 1970), evolutionary behavior (Moynihan 1958a & b, Beer 1975), physiology (Tucker 1972, Howell et al. 1974), foraging behavior (Andersson et al. 1981, Curtis et al. 1985, Greig 1984, Patton 1986), territoriality (Burger 1984), interspecific associations (Gotmark and Andersson 1980, Barnard and Thompson 1985), life history strategies (see Burger et al. 1980), and a number of other subjects. Because several gull species have been thoroughly studied, it is now possible to design interesting comparative studies dealing with ecology and behavior. Even with all the attention gulls have received from investigators, many unanswered questions remain. The papers and abstracts presented in this volume provide an outstanding indication of the directions gull research is taking and suggest approaches for further inquiry.

Twenty-one species of gulls breed in North America and three other species regularly visit the continent (Farrand 1983). Of the five genera involved, *Larus* includes the largest number of species (19). Both species of *Rissa* occur here and *Rhodostethia, Xema* and *Pagophila* each are represented by one species. Several species range widely over arctic waters or are nearly pelagic in the North Atlantic and Pacific oceans. Six of the 21 breeding species tend to nest in inland locations whereas 15 species are primarily coastal nesters. During the nonbreeding period, considerable overlap occurs in the ranges of the nonarctic species. The breeding ranges, however, are more distinct and only occasionally do more than two or three species share colony sites (Southern 1980, American Ornithologists' Union 1983).

Several species of gulls are good research sub-

[1] Department of Biological Sciences, Northern Illinois University, DeKalb, IL 60115.

jects because their colonies are relatively accessible and they nest in large numbers which allow investigators to obtain statistically important samples. During the last two decades alone, hundreds of papers have been published about gull migration and orientation, seasonal distribution, breeding biology, sex ratios, ecology, food habits and population size. Of the 21 species breeding in North America, 6 have received most of the research attention: Laughing Gull, *L. atricilla*; Ring-billed gull, *L. delawarensis*; California Gull, *L. californicus*; Herring Gull, *L. argentatus*; Western Gull, *L. occidentalis*; and Glaucous-winged Gull, *L. glaucescens*. Others such as the kittiwakes (*Rissa* spp.), have been studied thoroughly in the Old World. Considerably less is known about others (e.g., Franklin's Gull, *L. pipixcan*; Bonaparte's Gull, *L. philadelphia*; Mew Gull, *L. canus*; Iceland Gull, *L. glaucoides*; Ross' Gull, *Rhodostethia rosea*; Sabine's Gull, *Xema sabini*; and Ivory Gull, *Pagophila eburnea*).

Bent's (1947) "Life Histories" illustrates the nature of gull research prior to the middle of this century. Shortly thereafter, Tinbergen's (1953) classic study of the Herring Gull stimulated numerous ethological studies and field experiments. Moynihan (1958a & b) described the visual and auditory displays of several larid species and provided the types of information considered necessary for an ethogram. Such studies provided us with significant descriptive information but they also caused us to de-emphasize the importance of individual differences in behavior. The fixed action pattern concept of Lorenz (see translations, 1970) left the impression that much of bird behavior was inflexible. We now know that more plasticity exists in the performance of gull displays and the components of other behaviors than earlier investigators proclaimed. For example, gull chicks can stimulate adults to feed them by pecking at portions of the the bill other than the salient spot or ring that adults possess. Also, as parental care progresses during the nesting season, experienced parents may attempt to feed without any begging by the chick (Henderson 1975; pers. obs.). Experienced parents and chicks show more variability in the behaviors associated with parental care than do novice parents and their chicks. These raw materials for behavioral change are awaiting the influences of selective pressures and they should be catalogued by investigators (e.g., Hand 1979). Documenting the variability in behavior, rather than ignoring it in favor of the sample mean, may provide us with insight into the rate at which shifts in behavioral tendencies may occur.

Gull researchers have contributed to at least three recent findings that have influenced the way that avian field research is conducted. Researcher sensitivity to these factors will result in more accurate data collection and analysis, and conclusions that more correctly describe how a given species is performing. (1) Gull investigators are becoming increasingly cognizant of the importance of long-term studies (e.g., Mills 1973, Coulson and Thomas 1985), which take into account what happens throughout a particular breeding season as well as throughout the lifespan of individual gulls. This is particularly applicable in the case of studies dealing with population trends, reproductive success and habitat selection. (2) The project designs and methods used by many researchers clearly show that they are now cognizant of the effects of investigator-caused disturbance in gull colonies (Hunt 1972, Robert and Ralph 1975, Hand 1980, Fetterolf 1983). Ignoring these effects when designing or conducting a study can seriously bias the data collected, particularly in studies measuring chick survivorship, parental care, aggressiveness and territoriality. (3) Methods of marking gulls may influence the accuracy of data collected and seriously bias the outcome of a study. For example, Southern and Southern (1985) showed that wing markers detrimentally influence the breeding behavior of Ring-billed Gulls. Use of this marking method during studies dealing with mate fidelity, longevity, site tenacity or other studies requiring unimpeded return to the site of marking should be avoided. It is no longer possible for investigators to discount the possibility that their experimental methods may influence the behavior of their research subjects. Ways of avoiding such complications must be developed during the planning stages rather than attempting to work around them statistically during the analysis stage.

The topics covered by this volume are some of those having the greatest importance to larid researchers today. The papers and abstracts are grouped into five subject areas: life histories, behavior, foraging, habitat selection and hybridization. Information of these types is accumulating gradually for most gull species. Particular ones are more thoroughly studied than others but sufficient data exist for a comparative approach possibly relating the similarities and differences to morphological characteristics, ecological variables associated with differing geographical ranges, and the effects of sympatry. The recent work of Hoffman (1984) is an outstanding example of the value of the comparative approach. Components of life history and ecological characteristics of species are more difficult to describe quantitatively than skeletal features; however, someone needs to accept the challenge and synthesize the behavioral and ecological data for gulls, particularly sympatric species. Burger (1980) stands out as a major contributor of species-specific data as well as a synthesizer of interspecific strategies.

The 11 full-length papers in this volume are a significant contribution to gull biology. The abstracts describe studies we will learn more about in the months to come as the associated papers are published. Following are some of my reactions to these papers. The abstracts are not discussed because of space limitations and the inability of the reader to refer to the full paper for details.

The lead paper by Walter V. Reid examines factors that may limit clutch size in the Glaucous-winged Gull. As with most *Larus* gulls, the clutch size of this species usually is limited to 3 eggs, with 4 or more eggs being relatively infrequent, or associated with female-female pairs (Conover 1984). Several hypotheses have been presented to account for the high frequency of 3-egg clutches. The energetic cost of egg formation has been offered as one explanation for egg and clutch size in gulls (e.g., Boersma and Ryder 1983). Measuring weight gains or foraging success of individual gulls after they reach the breeding range may not be the best approach for examining this possibility, although it is regularly used. More important may be the body condition of females when they arrive on the breeding grounds. Not infrequently, gulls spend relatively little time foraging during the early stages of the nesting cycle (i.e., prelaying; pers. obs.). It appears, therefore, that fat reserves may not only contribute to survival at this time but may provide some of the energy required for egg production by early nesters. Ryder (pers. comm.) is investigating whether or not this may be the case for Ring-billed Gulls.

Reid suggested that the incubation capacity of gulls may impose an upper limit on clutch size. No evidence exists, however, to show that possession of only 3 brood patches prevents gulls from successfully incubating more than 3 eggs (Vermeer 1963, Coulter 1973), although Coulter (this symposium) showed that hatching success is highest for 3-egg clutches. The brood-rearing capability of parent gulls has been suggested as another factor possibly responsible for limiting clutch size (Haymes and Morris 1977), although some gulls are capable of rearing more than three young (e.g., Coulter, this symposium). In spite of this, average reproductive success seldom exceeds 1.5 chicks per pair (Blokpoel and Tessier 1986) and may be considerably lower. It is likely that no single factor is responsible for the prevalence of 3-egg clutches in gulls. The multiple hypothesis approach of Winkler (1985) shows the advantages of a broader perspective to questions such as this.

Reid also calls attention to the small c-egg (third laid) commonly reported for gulls and suggests that it may not represent an adaptation for brood reduction. Instead he considers it a non-adaptive consequence of energy shortages during laying. He also points out that asynchronous hatching in gulls may be an adaptation for maximal growth rather than an adaptation for food stress. The pattern of hatching in some gull species such as the Ring-bill, however, is variable with some clutches hatching all 3 eggs on the same day but hatching in others is spread over 2–6 days (Clark and Wilson 1981; Southern, in prep.). Reid's explanation, therefore, is not generally applicable to all gull species.

D. Michael Fry, C. Kuehler Toone, Steven M. Speich and R. John Peard examine the factors affecting skewed sex ratios in gulls, a subject that has received considerable attention during the last decade. Sex ratios skewed toward females are thought to result female-female pairs (Hunt and Hunt 1977, Ryder 1978, Ryder and Somppi 1979, Conover 1984). This phenomenon is indicated by the occurrence of supernormal clutches (SNC) and indexed by the SNC percentage within a colony. Causes of skewed sex ratios may be multifaceted as the authors describe. The finding that there is a decrease in the number of male gulls and an increase in the number of SNCs in areas polluted with organochlorines is extremely interesting. Once again we are reminded that all behavioral, morphological and physiological conditions we identify when examining large samples of organisms, as is possible in gull colonies, are not necessarily adaptive (Gould and Lewontin 1979, Hand 1979). Some, such as female-female pairing, may not be indicative of a new mode of parental care that can be expected to sweep through gull colonies, although some investigators seemed to imply this in the past (e.g., Hunt and Hunt 1977).

Egg predation by conspecifics is not uncommon when pair members are nesting asynchronously from most of the colony or when they are casual about attentiveness (pers. obs.). This is especially true of gulls with small nesting territories. Attentive behavior by both sexes of parents during incubation and early stages of chick development appears to be an effective defense against this form of predation (L. A. Hanners MS; Shugart and Fitch, abstract this symposium). Individual differences occur, however, in the performance of parental care by gulls and this may contribute to differential brood success. Ralph D. Morris examines time-partitioning of clutch and brood care activities as measures of parental quality in Herring Gulls. His findings confirm that pairs displaying the greatest synchrony in parental care produce the most young.

The subject of survivorship and mortality is

fundamental to understanding the dynamics of avian populations and associated life history strategies. According to Larry B. Spear, Harry R. Carter, Teresa M. Penniman, Jay F. Penniman and David G. Ainley, only four studies provide reliable information on survival rates in adult gulls. These authors also report finding no quantitative estimates of the various causes of mortality that affect gull age or sex composition. Their paper points to one of the areas of gull research that requires further attention by investigators. Especially needed are reliable techniques for predicting changes in gull populations on a regional basis and for cataloging the factors which limit population growth of these successful generalists.

Gull populations have increased dramatically across the Northern Hemisphere during recent decades thereby providing opportunities for investigations of the causes and effects of such changes. Conditions responsible for these significant population changes are not restricted to a single region nor to a single species. Interesting biological questions are associated with these population changes and the resulting inter- and intra-specific conflicts. Arie L. Spaans, Alle A. N. de Wit and Marianne van Vlaardingen examined the effects of increased population size on Herring Gull breeding success in The Netherlands. Between 1968 and 1984, the increase in Herring Gull population size was more than five-fold. In the authors' study plots, the increase was three-fold with a corresponding decrease in the number of young fledged per pair. Interestingly, under these conditions, experienced breeders were producing most of the offspring and the breeding schedule had advanced 4–9 days since the 1960s. Gulls are breeding earlier in other parts of the world as well. For example, since 1975 the onset of hatching of Ring-billed Gulls at Rogers City, Michigan, has advanced 7–10 days with the first chicks now hatching in mid-May (Southern, in prep.). It is possible that the factors associated with this shift involve more than density-dependent phenomena, as suggested by Spaans and his co-workers for Herring Gulls. Possibly subtle changes in temperate zone climatic conditions are having a gradual affect. Other circumstances such as rising Great Lakes and ocean water levels may be a further reflection of such changes.

The subject of parental recognition of their young has received the attention of several investigators working with various species of gulls (e.g., Tinbergen 1953; Beer 1970, 1979; Miller and Emlen 1975). Intuitively it would seem that ground nesting colonial gulls with potentially mobile young should possess some method for distinguishing their young from those of nearby conspecifics. At least this would be the case if natural selection was occurring at only the individual level and the concept of inclusive fitness was applicable. Although earlier studies produced evidence supportive of these contentions, the results from recent ones, including those of Joseph G. Galusha and Ronald L. Carter presented here, indicate that recognition may not be well perfected in gulls and that adoptions or temporary care of young other than a parent's own may occur (Holley 1981, 1984; Spear et al. 1986). This raises some interesting evolutionary questions, including the significance of unintended cooperation in breeding gulls. In studies without investigator or other disturbances, chick mortality often is not a consequence of chicks invading neighboring territories. Some adults show varying levels of tolerance or acceptance of chicks other than their own. The result often is temporary or permanent adoption (Southern, in prep.). Selective advantages to acceptance of chicks by neighbors could exist, particularly in the case of gulls with small territories. Our skepticisms about group selection should not close our minds to such possibilities as the benefits may be at the individual level. The conclusion of Galusha and Carter that adult gulls do not recognize their chicks individually but accept or reject them on the basis of their behavior deserves careful attention by other investigators. Short-term and long-term adoptions also occur regularly in Ring-billed Gulls (pers. obs.). A possibility worthy of testing is that acceptance of "foreign" chicks, particularly by experienced pairs that have lost their own chicks, contributes to colony stability during a particular nesting cycle by keeping more adults at the colony. If social facilitation has any importance to breeding gulls, particularly those with small territories, assuring an optimally sized social assemblage may be selectively advantageous.

As information about gull species increases, it becomes increasingly important to synthesize the data and present an overview of what is typical as well as what is unique to individual species or groups of species. Joanna Burger presents a paper that accomplishes this goal using data she collected for 15 species of gulls in North America, Africa, Australia and Europe. Few investigators have had such vast experience with the world's gull species. Although an assortment of authors cited by Burger have discussed the age-related differences in feeding ability, she is the first to use uniformly collected data to examine foraging efficiency for a large number of widely distributed gull species. Her results solidify the theory that delayed maturation is likely to occur in cases where foraging difficulties exist.

The responses of nesting gulls to nocturnal predators and the effects of predators on breeding success are subjects of broad interest to gull researchers (L. Southern et al. 1982). Joseph R. Jehl and Charles Chase III discuss the foraging patterns and prey selection of predators, especially Great Horned Owls (*Bubo virginianus*) on California Gulls. As in other studies (e.g., Southern et al. 1985), the authors found that adult gulls left the colony during owl attacks. As a result, indirect chick losses were a regular occurrence. The hunting patterns of owls were regular and predictable. Adult losses were low but chick losses occasionally were great. This study provides further evidence that the "antipredator" behavior of gulls, particularly under nocturnal conditions, is little more than avoidance by leaving when predators are present. If adults make any attempt to protect their offspring at night, it is ineffective against most persistent nocturnal predators (see Southern et al. 1982 for a review). Jehl and Chase also provide important information about who gets killed and why, which has implications for habitat selection and colony siting. Because the impact of predators can be local but severe, sampling methods in large colonies must be considered carefully.

Considerable attention is being directed at the topics of habitat and nest-site selection by gulls. Kees Vermeer and Kevin DeVito compare the characteristics of sites selected by Mew Gulls and Glaucous-winged Gulls. Information about the Mew Gull is especially interesting as this species has been little studied in North America. On Vancouver Island about 80% of the Mew Gulls nested as solitary pairs. Nest sites frequently were on the tops of poles or other objects which were surrounded by water. The Glaucous-winged Gull, on the other hand, is primarily a colonial nester. Interspecific plasticity in nest site selection by both species was noted.

Habitat selection has received considerable attention from gull biologists, and justifiably so (Bongiorno 1970, Burger and Shisler 1978, Erwin et al. 1981, Montevecchi 1978). A common flaw in many such studies, however, is that the investigator assumes that the conditions under which gulls may be nesting when a study starts are the same as those that existed when individual gulls first occupied the site. Changes in cover type and density may occur within a breeding season because of plant growth and even more dramatic changes may occur over the lifespan of individual gulls. Since nest site tenacity is well documented in gulls (L. Southern, in prep.), as is mate fidelity, the probability exists that given nest sites will change over time because of plant succession or other variables. Long-term studies are necessary to distinguish between the effects of nest site selection and effects associated with plant succession or other time-related factors (i.e., time vs. tradition) on an individual's total reproductive output. It appears that gulls continue to use sites long after the habitats that existed when they selected them no longer are evident. In this volume, Raymond Pierotti examines the behavioral consequences of habitat selection in Herring Gulls. He compares the time budgets, rates of aggressive behavior and diets of gulls nesting in three different habitats in Newfoundland. His results demonstrate that habitat choice may influence the type and frequency of particular behaviors which, in turn, influence reproductive success. Studies such as this which address the variability within a population or species are extremely important. Variation appears to be the rule rather than the exception, particularly when we are dealing with gulls because so many exhibit high levels of plasticity in behavior. From the evolutionary standpoint, tomorrow's trends exist in today's variability. It is well documented that behavioral changes can occur over relatively short spans of time. Devoting more attention to such things as how variability in habitat preference influences the production of offspring may give us a better record of evolution in progress.

By studying hybridization in nature, it is possible to assess the evolutionary status of closely related populations (Moore 1977). If members of two populations successfully and freely interbreed whenever their ranges overlap, taxonomists should seriously consider classifying them as conspecifics (Hoffman et al. 1978). Hybridization occurs between many of the large *Larus* gulls (Tinbergen 1953, Ingolfsson 1970, Jehl 1971). In this volume, Aonar Ingolfsson, who is recognized for his long-term studies of gulls in the far north, presents information collected over 15 years about the extensive hybridization between the Herring and Glaucous gulls in Iceland. Herring Gull-like birds raised fewer young per nesting attempt that more Glaucous Gull-like individuals. Birds of intermediate appearance had a higher incidence of non-breeding than the others. It appears that the population in this area is not becoming more Glaucous Gull-like, possibly as a result of continuing immigration of pure Herring Gulls from Europe.

A variety of topics is discussed in this volume. I am confident that you, the reader, will find them stimulating as well as a significant contribution to the gull literature. Ernst Mayr (1984) vividly portrayed the contributions ornithologists have made to biology. It is clear that we are continuing to make progress. Our understanding of the appropriateness of techniques, the importance of long-term studies, and our attention to the effects our own activities may be having on the accuracy of our data, will enable gull biologists to make even greater contributions in the future.

This volume is the first joint publication of the

Pacific Seabird Group and the Colonial Waterbird Group and originated at their First Joint Meeting. We hope this achievement will stimulate further cooperation between two organizations which together can have profound influence on colonial waterbird and seabird conservation and management in this hemisphere and worldwide.

LITERATURE CITED

AMERICAN ORNITHOLOGISTS' UNION. 1983. Check-list of North American birds. Sixth Edition. Allen Press, Lawrence, KS.

ANDERSSON, M., F. GOTMARK, AND C. WIKLUND. 1981. Food information in the Black-headed Gull, *Larus ridibundus.* Behav. Ecol. Sociobiol. 9:195–200.

BAERENDS, G. P., AND R. H. DRENT. 1970. The Herring Gull and its egg. Behav. Suppl. 17:1–312.

BARNARD, C. J., AND D. B. A. THOMPSON. 1985. Gulls and Plovers, the ecology and behaviour of mixed-species feeding groups. Columbia Univ. Press, New York. 302 pp.

BEER, C. G. 1970. On the response of Laughing Gull chicks (*Larus atricilla*) to the calls of adults. I. Recognition of the voices of parents. Anim. Behav. 18: 652–660.

BEER, C. G. 1975. Multiple functions and gull displays. Pp. 16–54, *in* G. P. Baerends, C. G. Beer, and A. Manning (eds.). Functions and evolution of behaviour. Clarendon Press, Oxford.

BENT, A. C. 1947. Life histories of North American gulls and terns. Dodd, Mead & Co., New York. 333 pp.

BLOKPOEL, H., AND G. D. TESSIER. 1986. The Ring-billed Gull in Ontario: a review of a new problem species. Occas. Paper No. 57, Canadian Wildlife Serv., Ottawa. 34 pp.

BOERSMA, D., AND J. P. RYDER. 1983. Reproductive performance and body condition of earlier and later nesting Ring-billed Gulls. J. Field Ornithol. 54:374–380.

BONGIORNO, S. 1970. Nest-site selection by Laughing Gulls (*Larus atricilla*). Anim. Behav. 18:434–444.

BURGER, J. 1980. The transition to independence and postfledging parental care in seabirds. Pp. 367–447, *in* J. Burger, B. L. Olla, and H. E. Winn (eds.). Behavior of marine animals, Vol. 4. Plenum Publ. Corp., New York.

BURGER, J. 1984. Pattern, mechanism, and adaptive significance of territoriality in Herring Gulls (*Larus argentatus*). Orn. Mono. No. 34, A.O.U., Washington, DC. 92 pp.

BURGER, J., B. L. OLLA, AND H. E. WINN (eds.). 1980. Behavior of marine animals, Vol. 4. Plenum Publ. Corp., New York.

BURGER, J., AND J. SHISLER. 1978. Nest site selection and competitive interaction of Herring and Laughing gulls in New Jersey. Auk 95:252–266.

CLARK, A. B., AND D. S. WILSON. 1981. Avian breeding adaptations: Hatching asynchrony, brood reduction, and nest failure. Quart. Rev. Biol. 56:253–277.

CONOVER, M. R. 1984. Occurrence of supernormal clutches in the Laridae. Wilson Bull. 96:249–267.

COULSON, J. C., AND C. S. THOMAS. 1985. Changes in the biology of the kittiwake *Rissa tridactyla*: a 31-year study of a breeding colony. J. Anim. Ecol. 54: 9–26.

COULTER, M. C. 1973. Clutch size in the Western Gull. Ibis 115:474.

CURTIS, D. J., C. G. GALBRAITH, J. C. SMYTH, AND D. B. A. THOMPSON. 1985. Seasonal variations in prey selection by estuarine Black-headed Gulls (*Larus ridibundus*). Est. Coast. Shelf Sci. 21:75–89.

DRURY, W. H., AND J. A. KADLEC. 1974. The current status of the Herring Gull population in the northeastern United States. Bird-Banding 45:297–306.

ERWIN, R. M., J. GALLI, AND J. BURGER. 1981. Colony site dynamics and habitat use in Atlantic coast seabirds. Auk 98:550–561.

FARRAND, J., JR. 1983. The Audubon Society master guide to birding. Vol. 2, Gulls to dippers. A. A. Knopf, New York. 398 pp.

FETTEROLF, P. M. 1983. Effects of investigator activity on Ring-billed Gull behavior and reproductive performance. Wilson Bull. 95:23–41.

GOTMARK, F., AND M. ANDERSSON. 1980. Breeding associates between Common Gull (*Larus canus*) and Arctic Skua (*Stercorarius parasiticus*). Ornis Scand. 11:121–124.

GOULD, S. J., AND R. C. LEWONTIN. 1979. The spandrels of San Marco and the Panglossian paradigm: a critique of the adaptationist programme. Proc. R. Soc. London B 205:581–598.

GRAHAM, F., JR. 1975. Gulls, a social history. Random House, New York. 179 pp.

GREIG, S. A. 1984. The feeding behaviour of *Larus argentatus* and other *Larus* gulls at refuse tips. Unpubl. Ph.D. thesis, Univ. of Durham.

HAND, J. L. 1979. Vocal communication of the Western Gull (*Larus occidentalis*). Unpubl. Ph.D. dissertation, Univ. California, Los Angeles, CA.

HAND, J. L. 1980. Human disturbance in Western Gull *Larus occidentalis livens* colonies and possible amplification of intraspecific predation. Biol. Conserv. 18:59–63.

HENDERSON, B. A. 1975. Role of the chicks' begging behavior in the regulation of parental feeding behavior of *Larus glaucescens.* Condor 77:488–492.

HOFFMAN, W. 1984. Phylogeny, feeding behavior, and wing structure in gulls, terns, and allies (Laroidea). Unpubl. Ph.D. dissertation, Univ. South Florida, Tampa, FL.

HOFFMAN, W., J. A. WIENS, AND J. M SCOTT. 1978. Hybridization between gulls (*Larus glaucescens* and *L. occidentalis*) in the Pacific northwest. Auk 95: 441–458.

HOLLEY, A. J. F. 1981. Naturally arising adoption in the Herring Gull. Anim. Behav. 29:302–303.

HOLLEY, A. J. F. 1984. Adoption, parent-chick recognition and maladaption in the Herring Gull (*Larus argentatus*). Z. Tierpsychol. 64:9–14.

HOWELL, T. R., B. ARAYA, AND W. R. MILLIE. 1974. Breeding biology of the Gray Gull, *Larus modestus.* Univ. Calif. Publ. Zool. 104:1–57.

HUNT, G. L., JR. 1972. Influence of food distribution and human disturbance on the reproductive success of Herring Gulls. Ecology 53:1051.

HUNT, G. L., AND M. W. HUNT. 1977. Female-female pairing in Western Gulls (*Larus occidentalis*) in southern California. Science 196:1466–1467.

INGOLFSSON, A. 1970. Hybridization of Glaucous Gulls *Larus hyperboreus* and Herring Gulls *L. argentatus* in Iceland. Ibis 112:340–362.

JEHL, J. R. 1971. A hybrid Glaucous × Herring gull from San Diego. Calif. Birds 2:27–32.

LORENZ, K. 1970. Studies in animal and human behaviour, Vols. 1 & 2. Harvard Univ. Press, Cambridge, MA. 401 & 366 pp.

LUDWIG, J. P. 1974. Recent changes in the Ring-billed Gull population and biology in the Laurentian Great Lakes. Auk 91:575–594.

MAYR, E. 1984. The contributions of ornithology to biology. BioSci. 34:250–254.

MILLER, D. E., AND J. T. EMLEN. 1975. Individual chick recognition and family integrity in the Ring-billed Gull. Behaviour 52:122–144.

MILLS, J. A. 1973. The influence of age and pair-bond on the breeding biology of the Red-billed Gull *Larus novaehollandiae scopulinus*. J. Anim. Ecol. 42: 147.

MONTEVECCHI, W. 1978. Nest site selection and its survival value among Laughing Gulls. Behav. Ecol. Sociobiol. 4:143–161.

MOORE, W. S. 1977. An evaluation of narrow hybrid zones in vertebrates. Quart. Rev. Biol. 52:263–277.

MOYNIHAN, M. 1958a. Notes on the behavior of some North American gulls: II. Non-aerial hostile behavior of adults. Behaviour 12:95–182.

MOYNIHAN, M. 1958b. Notes on the behavior of some North American gulls: III. Pairing behavior. Behaviour 13:112–130.

PATTON, S. R. 1986. Comparative foraging ecology of three species of gulls (*Larus*) at urban landfills in west-central Florida. Unpubl. Ph.D. dissertation, Univ. South Florida, Tampa, FL.

ROBERT, H. C., AND C. J. RALPH. 1975. Effects of human disturbance on the breeding success of gulls. Condor 77:495–499.

RYDER, J. P. 1978. Possible origins and adaptive value of female-female pairing in gulls. Proc. Colonial Waterbird Group 2:534.

RYDER, J. P., AND P. L. SOMPPI. 1979. Female-female pairing in Ring-billed Gulls. Auk 96:1–5.

SOUTHERN, L. K., S. R. PATTON, AND W. E. SOUTHERN. 1982. Nocturnal predation on *Larus* gulls. Colonial Waterbirds 5:169–172.

SOUTHERN, W. E. 1980. Comparative distribution and orientation of North American gulls. Pp. 449–498 *in* J. Burger, B. L. Olla, and H. E. Winn (eds.). Behavior of marine animals, Vol. 4. Plenum Publ. Corp., NY.

SOUTHERN, W. E., S. R. PATTON, L. K. SOUTHERN, AND L. A. HANNERS. 1985. Effects of nine years of fox predation on two species of breeding gulls. Auk 102: 827–833.

SPEAR, L. B., D. G. AINLEY, AND R. P. HENDERSON. 1986. Postfledgling parental care in the Western Gull. Condor 88:194–199.

TINBERGEN, N. 1953. The Herring Gull's world. Collins, London. 255 pp.

TUCKER, V. A. 1972. Metabolism during flight in the Laughing Gull (*Larus atricilla*). Am. J. Physiol. 222: 237–245.

VERMEER, K. 1963. The breeding ecology of the Glaucous-winged Gull (*Larus glaucescens*) on Mandarte Island, B.C. Occas. Papers B.C. Prov. Mus. 13:1–104.

WINKLER, D. W. 1985. Factors determining a clutch size reduction in California Gulls (*Larus californicus*): a multi-hypothesis approach. Evolution 39:667–677.

Studies in Avian Biology No. 10:8–25, 1987.

CONSTRAINTS ON CLUTCH SIZE IN THE GLAUCOUS-WINGED GULL

WALTER V. REID[1]

ABSTRACT.—I examined three factors that may limit the clutch size of the Glaucous-winged Gull (*Larus glaucescens*) to three or fewer eggs: the energetic cost of egg formation, the shelf-life of eggs, and the incubation capacity of adults. Incubation capacity was found to have a significant effect on the success of large clutches but it cannot explain the absence of 4-egg clutches. Energetic limitation following the initiation of the clutch may be a more important factor limiting clutch size to three.

I examined several aspects of the brood reduction hypothesis to determine whether the presence of brood reduction adaptations is compatible with evidence that clutch size is not limited by the brood rearing capacity of the adults. Asynchronous hatching was found to be beneficial regardless of the number of young that could be raised and thus is consistent with evidence that brood-rearing capacity does not limit clutch size. The small size of the third egg, generally considered to be another brood reduction adaptation, was found to be a result of energetic shortages during laying and thus may not be an adaptive mechanism for brood reduction.

Members of the family Laridae exhibit modal clutch sizes ranging from 1 to 3 eggs, 4-egg clutches occurring infrequently (see Conover 1984). Some of the 4-egg clutches reported are produced by female-female pairs (Conover 1984). Clutch-size distributions with modal and maximal values of 3 eggs are found in at least 11 larid species: the Herring Gull (*Larus argentatus*), Laughing Gull (*L. atricilla*), Glaucous-winged Gull (*L. glaucescens*), Common Tern (*Sterna hirundo*), California Gull (*L. californicus*), Lesser Black-backed Gull (*L. fuscus*), Black-headed Gull (*L. ridibundus*), Common Gull (*L. canus*), Western Gull (*L. occidentalis*), Ring-billed Gull (*L. delawarensis*), and Great Black-headed Gull (*L. ichthyaetus*) (Samorodov and Ryabov 1969, Conover 1984). This group encompasses a diverse range of both body sizes and feeding habits, though most species are relatively large and all breed in temperate zones. The reason for the truncation of the clutch size frequency distribution at three eggs is not clear. Three factors—the energetics of egg formation, incubation capacity, and brood rearing capacity—have received attention as factors potentially limiting clutch size to three or fewer eggs.

The energetic cost of egg formation is thought to explain patterns of variation in larid egg size, clutch size, and nesting phenology (Nisbet 1973, 1977, Mills 1979, Pierotti 1982, Schreiber et al. 1979, Mills and Shaw 1980, Boersma and Ryder 1983, Houston et al. 1983, Winkler 1983, 1985, Pierotti and Bellrose 1986). The energetic cost of egg formation, however, does not place a strict upper limit on egg production at 3 eggs because protracted laying can be induced in at least 4 of the 11 species exhibiting a truncated clutch-size frequency distribution (Herring Gull: Paludan 1951, Harris 1964, Parsons 1976, Pierotti 1982; Glaucous-winged Gull: this study; California Gull: Winkler 1983, 1985; Black-headed Gull: Weidmann 1956).

Similarly, the incubation capacity of gulls and terns may not impose a fixed upper limit on clutch size. Most large gulls have three brood patches (see Table 10) and it is possible that this brood-patch configuration results in a 3-egg limit (Vermeer 1963, Pierotti and Bellrose 1986). There are no studies, however, that support this hypothesis. Experimental manipulation of clutch size during incubation has shown that more chicks hatch from artificially enlarged clutches than from 3-egg clutches (Coulter 1973a, b).

There is also no evidence that clutch size is limited to 3 or fewer eggs by the brood-rearing capacity of the adults. In at least 4 species, adults are capable of rearing more than 3 young (Herring Gull: Haymes and Morris 1977; Glaucous-winged Gull: Vermeer 1963, Ward 1973; Lesser Black-backed Gull: Harris and Plumb 1965; Western Gull: Coulter 1973b).

To further complicate the question of clutch-size determination, many larids possess traits that potentially conflict with the observation that gulls are capable of raising more than 3 young. Two traits characteristic of all larids with 3-egg clutches are the presence of asynchronous hatching, and size reduction of the third (c-) egg relative to the first 2 (a- and b-) eggs. These traits are frequently cited as evidence supporting the brood reduction hypothesis (Lack 1968, O'Connor 1978, Clark and Wilson 1981, Hahn 1981, Slagsvold et al. 1984). Asynchronous hatching and the small size of the c-egg place the third chick at a disadvantage with respect to its siblings in competition for food. During years of food shortage, these traits are thought to facilitate the early mortality of chicks that could not be raised, thereby increasing food available to the surviving chicks. The presence of brood reduction traits is interpreted as circumstantial evidence that

[1] Department of Zoology NJ-15, University of Washington, Seattle, Washington 98195.

brood size is close to the limit set by food in years with high food availability (Lack 1968).

For two reasons, the presence of a brood reduction strategy may be compatible with the observation that gulls are capable of raising more than 3 chicks. First, those pairs capable of raising more than 3 chicks may not exhibit brood reduction adaptations. Second, even if all pairs do exhibit brood reduction adaptations, these adaptations are incompatible with the ability of the birds to raise more than 3 young only if they represent a cost during years when all young can be raised. For example, consider a pair that is capable of raising 4 young in a good year and 2 young in a bad year but lays a clutch of only 3 eggs. If the brood reduction strategy provides a benefit in bad years without reducing success in good years then there would be selection for the strategy regardless of the number of chicks that could be raised.

There is evidence, however, that brood reduction adaptations, particularly the small c-egg, do represent a cost in good years. The probability of chick survival was significantly correlated with egg size, controlling for order of laying, in the Common Tern (Nisbet 1973), the Herring Gull (Parsons 1970, 1975a) and the Black-headed Gull (Lundberg and Vaisanen 1979). Reduction in the size of the c-egg thus appears to reduce the probability of the survival of the third chick under any conditions. Consequently, the brood reduction strategy may conflict with evidence suggesting that clutch size is not limited during the chick stage.

In this paper I address two questions: first, why do Glaucous-winged Gulls have a modal clutch of 3, and second, why is the distribution truncated at 3 eggs. Because of evidence suggesting that clutch size is not limited by parental feeding ability, I focus on factors acting during laying and incubation that may limit clutch size to 3. In addition, because of the potential conflict between the presence of brood reduction adaptations and the assumption that clutch size is not limited during the chick period, I also examine the brood reduction hypothesis from the perspective of clutch size regulation.

I examined 3 factors potentially influencing the modal clutch size and the limitation of clutch size to 3 eggs. First, the energetic cost of egg formation could contribute to a 3-egg limit. Though in several species of gulls it has been demonstrated that females are capable of laying more than 3 eggs, the cost to the female of production of a fourth egg may be high enough that the net benefit of the egg is small. Second, the viability of unincubated eggs (shelf-life) may limit clutch size to 3. Most larids lay eggs at 2-day intervals and incubation does not begin until the b-egg is laid. Since incubation begins prior to the completion of the clutch, the first 2 eggs hatch 1 or 2 days before the third. While it is generally assumed that asynchronous hatching, and thus the timing of the onset of incubation, are adaptations for unpredictable food resources, the same pattern could result if the timing of the onset of incubation was dictated by a short shelf-life of unincubated eggs. If early onset of incubation was required for egg survival this could limit clutch size to 3 since a fourth chick would hatch nearly 4 days after the first and thus be at an extreme competitive disadvantage. Finally, the incubation capacity of adults places a proximate limit on the number of eggs that can be hatched. I evaluated the role of incubation capacity as a constraint on clutch size by measuring the hatching success of artificially enlarged clutches.

I examined 3 aspects of the brood reduction hypothesis to determine whether it conflicted with the assumption that clutch size is not limited during the chick-rearing stage. First, the conflict would be avoided if some pairs did not show evidence of brood reduction adaptations. I measured the natural patterns of hatching synchrony and c-egg size to determine whether the survival probability of chicks was equalized in some broods through synchronous hatching and uniform egg size. Second, I examined the costs and benefits of asynchronous hatching to determine whether predictions of the brood reduction hypothesis are met and to determine whether asynchronous hatching represents a cost under conditions where 3 or more young can be fledged. Finally, I tested the role of energetic limitations as an alternative explanation for the small c-egg. The reduced c-egg size appears to represent a cost to the adults under circumstances when 3 or more young can be fledged. If, however, the small size of this egg is not an adaptation for brood reduction, then it would be compatible with the assumption that clutch size is not limited during the chick period.

METHODS

GENERAL

This study was conducted on Protection Island, Washington (48°07′N, 122°55′W) between May and August of 1983–1985. All experiments were carried out on a 700 × 100-m sandspit which is used by roughly 5000 pairs of breeding gulls. Vegetation on the spit ranges from bare ground to 1.5-m tall grass (*Elymus mollis*). Chicks are fed almost exclusively fish, primarily sandlance (*Ammodytes hexapterus*) and herring (*Clupea harengus*), but adults forage both on natural food sources and at garbage dumps (10 to 25 km from the colony) throughout the breeding season.

In several experiments, I made use of data collected at 250 nests which had been monitored since 1983. At least one adult at each nest was color-banded. The banded adults were nest-trapped during incubation (see Amlaner et al. 1978), weighed, and measured (tarsus, culmen, bill depth, bill width, wing chord, radius, body length). Sex was determined for 83 birds by the observation of copulations, and for the remainder of the birds through the use of a discriminant function created for the birds of known sex (98.9% accuracy for birds of known sex). Weights and measurements of birds were log-transformed prior to all analyses.

All experiments were performed at nests sampled randomly with respect to the time of laying. Nests utilized in the study were marked and assigned to experimental groups prior to laying. The age of adults is known to influence laying phenology and clutch size in several larid species (Coulson 1963, Davis 1975, Coulson and Horobin 1976, Mills 1979, Mills and Shaw 1980). The ages of adults in experimental groups in this study represent a random sample of the birds present.

All significance tests are one-tailed unless otherwise noted.

Energetic Requirements

Food supplement

Forty nest scrapes or obvious territories in an area of dense vegetation were marked on 7 May 1985. On 23 days between 8 May and 5 June, approximately 200 g (dry weight) of a moistened mixture of Purina Cat Chow and Darigold Cat Food was placed beside odd numbered nests with even numbered nests serving as controls. The experiment was conducted in tall grass in order to minimize disturbance by crows. The food was placed in small containers and these were partially concealed in the grass adjacent to the nest scrape. Food was provided at each nest until the laying date of the a-egg. No eggs were laid at 8 marked territories, leaving a sample of 18 experimental and 14 control nests. Nests were checked daily until 22 May (date of first clutch initiation) and twice daily subsequently. I was absent from the island on two occasions for 3- and 4-day periods; laying dates during these periods were estimated to be 48 h prior to the laying of the b-egg (Vermeer 1963). Egg length and breadth were measured to the nearest 0.1 mm and the eggs were weighed to the nearest 0.5 g.

At an additional seven nests I provided food to pairs in which the females had been color-banded and for which I had obtained information on egg size and laying date in 1984. Other nests with marked females served as controls for these 7 pairs. For comparisons between these groups I used egg volumes calculated using the formula: volume = 0.476 × length × breadth.[2]

It was not possible to observe whether birds in the 40-nest grid ate the supplemental food. All birds at the 7 nests with banded birds were seen to eat the food, some within seconds of my departure from the territory.

Egg removal

At 16 of the nests in the 40-nest feeding grid and at additional 34 nests with 1 or more color-banded adults, I removed the a-egg within 12 h of laying in order to stimulate production of a fourth (d-) egg.

Shelf-life

Between 25 May and 13 June 1985, I collected the first egg from 113 clutches within 12 h of laying, marked each egg with the date (written on tape), and placed each in an artificial nest, composed of the lining of several nests, which I shaded and fenced to exclude predators. Air temperatures during this period ranged from 4–26°C. After leaving each egg unincubated for 0–8 days (2-day intervals) I substituted 2 or 3 of these experimental eggs for eggs in 34 clutches at marked nests which had been completed within the previous 24 h. The 34 nests were checked daily during hatching and the success of each egg was recorded.

Incubation Capacity

Between 25 May and 13 June 1985, I manipulated the clutch size of 89 completed nests by adding or removing between 1 and 3 eggs. Manipulations were done within 4 days of clutch completion and eggs that were added to nests were of the same age as the eggs already present in the nest. I did not switch eggs between control clutches of 3 eggs. All nests were checked daily during hatching and hatching success was recorded.

Size of C-egg

Between 25 May and 13 June 1985, supplemental food was provided (as above) to 31 nests within 24 h of the laying of the a-egg and on each of the subsequent 4 to 5 days. Twelve nests were excluded from the analysis because of egg loss prior to weighing (n = 6) or because the completed clutch contained fewer than 3 eggs (n = 6). Each egg was measured and weighed within 48 h of laying. Eggs of known laying sequence in the remainder of the colony served as controls.

To examine patterns of attendance at the nest during and prior to incubation, I observed 87 nests from 3 elevated (2 m) wooden blinds. I conducted 15 3-h watches between 24 May and 30 June 1985. Each nest was observed on an average of 3.6 occasions. At 10-min intervals I scanned all nests visible from the blind and recorded the presence or absence of each member of the pair. At all but 8 of the nests at least 1 bird was color-banded. The importance of time budget information on the day of laying of the a-egg was recognized late in the season and so the 8 unbanded pairs, without eggs, were chosen and followed through egg laying. The median laying date for nests observed during the laying of the first egg (14 June) was later than for the colony as a whole (3 June), and there is a potential that this may have introduced some variance into the measured attendance patterns. I report attendance patterns only in terms of the amount of time both members of the pair were present since this removes the potential error of misidentification of the bird.

Hatching Synchrony

Natural pattern

Fifty-four nests, chosen randomly from the entire sample of 300 nests followed in 1985, were checked twice daily at the time of hatching to determine the time span between the hatching of the first and third chick. The order of laying was not known for all eggs so I could not calculate the relative size of the c-egg. Instead, I calculated the ratio of the smallest to largest

TABLE 1

CLUTCH SIZE DISTRIBUTIONS OF NESTS WITH AND WITHOUT SUPPLEMENTAL FOOD PRIOR TO LAYING

Group	Food supple- ment	Number of eggs laid 1	2	3	4	Total nests
First egg not removed						
Experimental	Yes	0	2	6	0	8
Control	No	0	2	6	0	8
First egg removed						
Experimental	Yes	2[a]	2	4	2	10
Control	No	0	1	4	1	6
Banded control	No	1[a]	3	13	17	34

[a] Nests abandoned after removal of egg.

egg in each clutch as a measure of the size range of eggs; this ratio should estimate the relative size of the c-egg.

Manipulations

Between 25 May and 13 June 1985, hatching synchrony was manipulated at 46 nests within 10 days of clutch completion. Under normal conditions, the third chick hatches 30 h after the second (see below). At 22 "synchronous" nests the c-egg was exchanged with a c-egg laid 1 day previously to create clutches in which the b- and c-eggs hatched synchronously. At 24 "asynchronous" nests the a-egg was exchanged with an a-egg laid 2 days previously to create a pattern of hatching that would result if incubation began on the day of clutch initiation. Another 31 nests that hatched all 3 chicks were not manipulated and served as controls. For several reasons (egg death, predation, and chick death prior to the hatching of all 3 chicks), only 11 of the synchronous and 9 of the asynchronous nests in the original design could be used in the experiment. Consequently, I created 32 additional experimental broods by adding 1 newly hatched (wet) chick to each of 5 of the original synchronous nests and 11 of the asynchronous nests that had lost 1 egg, and by replacing young chicks (less than 3 day old) at 16 other nests with 3 newly hatched chicks. The hatching order of the chicks added to these nests was not known. Results from the entire sample of nests matched the results from nests in the original study design and in the following analysis only the results for the entire sample of 32 synchronous and 20 asynchronous nests are reported.

Chicks at each nest were individually marked with tape bands on hatching and banded with aluminum bands on day 20. At approximately day 0, 10, 20, and 35 the chicks were weighed and the tarsus, culmen, and (on day 20 and 35) wing chord were measured. Weights and measurements were log-transformed prior to analysis. Chicks that were seen after day 32 are assumed to have fledged (fledging does not actually occur until approximately day 40). Sixty of 85 chicks that did not fledge were found dead and the age of death was estimated to the nearest day. The remainder of the chicks are presumed to have died and age of death was taken to be the age when last observed (19 of the 25 were not found at the 10-day check).

Growth rates were compared by testing for size differences at age 20 and 35. Two measures of size were used: chick weight, and a principal component factor score ("chick size") combining all measurements. Because of the substantial variation in chick weight resulting from periodic feedings, body size measurements are preferable indicators of growth. Principal components analysis allows the incorporation of several measurements of the size of the chick into 1 score reflecting overall size. Separate principal components analyses were performed for chicks of age 17–24 and age 32–37. Loadings on PCl at age 20 were: culmen .87, tarsus .89, weight .95, wing .86; and at age 35: culmen .83, tarsus .83, weight .96, wing .79. Not all chicks were measured at exactly ages 20 and 35. I adjusted the measured weights and sizes to these ages using the slopes of regressions of weight and size on age for the 2 intervals of 17–24 days and 32–37 days. Mean chick weights and chick sizes were then calculated for each nest (to avoid violation of the assumption of independence of measurements).

RESULTS

ENERGETIC REQUIREMENTS

Timing of laying

There was no significant difference in the timing of clutch initiation between food supplemented (n = 18) and control nests (n = 14) (median laying dates were 2 June and 3 June respectively; Mann-Whitney U, $P > .10$). Food had been provided for 13 days prior to the initiation of the first clutch (22 May). The seven

TABLE 2

FACTORS INFLUENCING THE TENDENCY FOR BIRDS TO LAY A FOURTH EGG FOLLOWING REMOVAL OF THE FIRST EGG: MEAN ± SD (N)

Factor	Number of eggs laid: Three or fewer	Four	Significance[a]
Weight of a-egg	96.2 ± 6.6 (16)	93.0 ± 8.1 (16)	.21
Laying date of a-egg (days after 1 May 1985)	34.2 ± 4.7 (17)	31.2 ± 4.2 (16)	.04

[a] Two-tailed Mann-Whitney U.

TABLE 3
SHELF LIFE OF UNINCUBATED EGGS

	Days without incubation				
	0	2	4	6	8
Initial number of eggs	23	22	24	21	23
Total hatching (%)	20 (87)	20 (91)	17 (71)	17 (81)	16 (70)
Total lost during incubation	2	0	5	3	4
Percent success of eggs not lost	95	91	89	94	84

food supplemented nests with banded females initiated clutches 2.7 (SD = 2.1) days earlier in 1985 than in 1984; 71 control nests initiated clutches 1.9 (SD = 6.7) days earlier (Mann-Whitney U, $P > .20$).

Egg size

There was no difference in the weight of the a-egg between food supplemented (95.7 g, SD = 6.2, n = 18) and control nests (95.0 g, SD = 8.6, n = 14) in the 40-nest grid (t = .26, $P > .25$). There was also no difference in a-egg weight between all supplemented nests (96.4 g, SD = 7.1, n = 25) and 131 nests in the remainder of the colony for which I had accurate weights of the a-egg (95.0 g, SD = 7.8) (t = .87, $P > .10$). There was no difference in the change in total clutch volume between 1984 and 1985 when the 7 supplemented nests with records of egg size in 1984 were compared to 71 control nests (supplemented: +1.85 cc, SD = 2.01; control: +.89 cc, SD = 3.73; Mann-Whitney U, $P > .10$).

Egg removals

A fourth egg was laid in 40% of nests from which the first egg was removed (Table 1). There was no indication that birds at nests which had received supplemental food were more likely to lay a fourth egg. Pairs in the 40-nest feeding grid (food supplemented and control combined) were less likely to produce a d-egg than pairs at the 32 nests with banded birds (G = 4.72, df = 1, $P < .05$). The reason for this difference is not clear, though it may be due to the greater disturbance caused by my regular feeding visits to the 40-nest grid.

If female condition influences the ability to lay a d-egg, then it would be predicted that females laying large eggs would be more likely to lay a d-egg. There was no relationship between the size of the a-egg and the tendency to lay a d-egg (Table 2). Birds that laid a d-egg, however, initated clutches on average 3 days earlier than those that did not (Table 2).

SHELF-LIFE

Twenty-four of the 113 eggs involved in the shelf-life experiment did not survive to hatch. Hatching success was not affected by the amount of time that the eggs were unincubated (Table 3; G = 5.28, $P > .25$). Thirteen of the eggs that did not survive were lost from the nest prior to hatching. I also examined the hatching success of only those eggs that were present in the nest after the standard incubation period and again there were no differences among groups (Table 3; G = 2.49, $P > .50$).

TABLE 4
EFFECT OF CLUTCH SIZE ON HATCHING SUCCESS[a]

	Clutch size				
	1	2	3	4	5
Number of nests	18	14	19	20	18
Number of eggs	18	28	57	80	90
Eggs developed (%)	13 (72)	24 (86)	51 (89)	49 (61)	60 (67)
Eggs hatched (%)	13 (72)	24 (86)	50 (88)	47 (59)	48 (53)
Eggs developed per nest	.72	1.71	2.68	2.45	3.33
Egg hatched per nest	.72	1.71	2.63	2.35	2.67
Number of nests hatching one or more (%)	13 (72)	13 (93)	19 (100)	16 (80)	13 (72)

[a] Highest value underlined.

TABLE 5
VOLUME (CC) OF EGGS OF THE GLAUCOUS-WINGED GULL BY SEQUENCE OF LAYING: MEAN ± SD (N)[a]

Year	Egg order		
	First	Second	Third
Three-egg clutches			
1983	86.35 ± 6.22 (89)	85.41 ± 6.19 (62)	79.79 ± 5.97** (82)
1984	84.79 ± 6.66 (88)	83.34 ± 6.81 (47)	77.52 ± 6.76** (92)
1985	85.96 ± 7.49 (72)	84.64 ± 8.14 (43)	79.75 ± 7.18** (73)
Combined	85.69 ± 6.77 (249)	84.64 ± 6.99 (152)	78.93 ± 6.70** (247)
Two-egg clutches			
1983	83.95 ± 6.71 (20)	82.64 ± 7.39 (18)	
1984	84.63 ± 8.17 (30)	81.37 ± 7.36 (29)	
1985	84.59 ± 5.91 (48)	81.20 ± 5.86* (35)	
Combined	84.47 ± 6.77 (98)	81.58 ± 6.70* (82)	

[a] Difference between each egg and the first egg is tested.
* $P < .05$, two-tailed t-test.
** $P < .001$.

INCUBATION CAPACITY

Hatching success differed significantly among clutches of different size (Table 4; $G = 28.0$, $df = 4$, $P < .001$). Peak hatching success (88%) was found for clutches of 3 eggs, and success fell rapidly in larger clutches. Part of the decline in hatching success in large clutches could be attributed to the tendency for pairs to stop incubation of viable (and sometimes pipped) eggs after 3 or 4 chicks had hatched. Consequently I also present results for 'development success', that is, the percent of eggs for each clutch size which developed to the point of pipping.

The average number of eggs hatched per nest was highest for clutches of 3 and 5 eggs; however, differences among clutches of 3 or more eggs were not significant (Kruskal-Wallis ANOVA, $P = .54$). The average number of developed eggs per nest differed significantly among clutches of 3 or more eggs (Kruskal-Wallis, $P = .05$). The number of developed eggs per nest was significantly higher in clutches of 5 than in clutches of 3 eggs (Mann-Whitney U, two-tailed $P = .04$). The probability of hatching at least 1 chick was highest in clutches of 3.

SIZE OF THE C-EGG

In each of the 3 years of this study the c-egg was significantly smaller than the a-egg (Table 5). In 2-egg clutches the b-egg was smaller than the a-egg only in 1985. Food supplementation provided on the day of laying of the a-egg resulted in an increase in the size of the c-egg (Table 6). The size of the c-egg in supplemented nests did

TABLE 6
EFFECT OF FOOD SUPPLEMENTATION ON THE WEIGHT OF THE C-EGG: MEAN (G) ± SD (N)[a]

	Egg order		
	First	Second	Third
Control	95.74 ± 8.27 (72)	94.56 ± 8.89 (43)	89.15 ± 8.23 (71)
Food supplement	94.95 ± 6.48 (19)	95.39 ± 7.46 (19)	93.06 ± 8.54* (18)[b]

[a] Differences between eggs of same order in laying sequence are tested.
[b] One egg was broken.
* $P < .05$, one-tailed t-test.

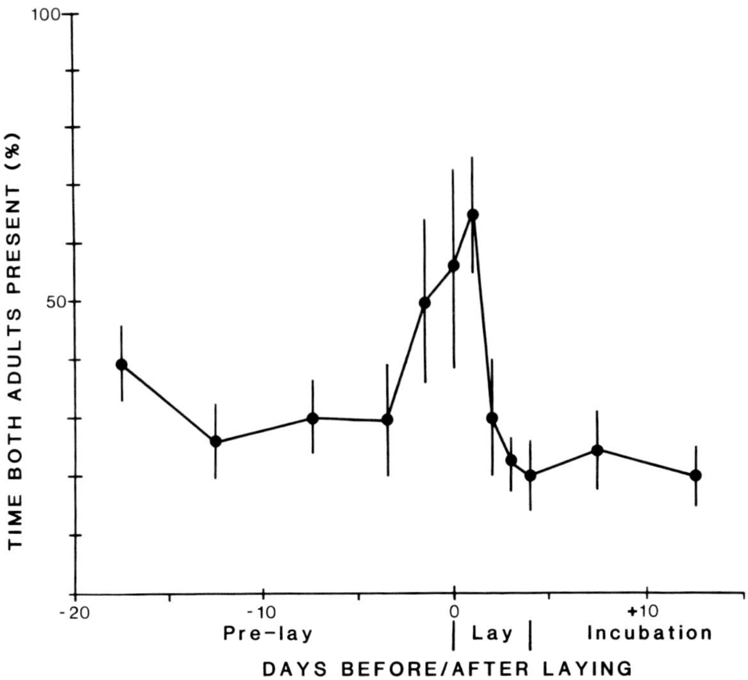

FIGURE 1. Percent of time during which both adults were present on territory as a function of days before and after laying of a-egg (Day 0). Means calculated from an average of 19.3 different nests (9.9 nests between −4 and +4 days). Mean ± SE.

not differ from the size of the a-egg (t = .75, $P > .10$; a difference less than 4.23 g could not be detected with this test).

The amount of time during which both members of a pair were on territory increased immediately prior to the initiation of laying and declined thereafter (Fig. 1).

Synchrony

At 54 nests that were checked twice daily, the third chick hatched 41.2 h (SD = 16.8, range 12–72 h, n = 54) after the first chick. The second chick hatched 9.7 h (SD = 9.1, range 0 to 36 h, n = 48) after the first. There was a significant positive correlation between the date of hatching and the length of time between the hatching of the first and third chicks (r = .27, $P = .03$, Spearman rank), but not with either egg size (r = −.01, $P > .40$) or the range of egg sizes in the clutch (r = −.01, $P > .40$).

Among nests used in the experimental study of synchrony (checked daily during hatching) the interval between hatching of first and third chicks was 9.4 h for synchronous (SD = 17.3, range: 0–48 h, n = 32), 39.1 h for control (SD = 15.1, range: 0–48 h, n = 31), and 88.8 h for asynchronous nests (SD = 25.9, range: 48–144 h, n = 20).

There were no significant differences in the number of chicks raised to day 35 between the experimental groups (Table 7; G-tests between each pair, $P > .25$). The success of asynchronous nests, however, is artificially inflated because I did not include nests that failed to hatch all 3 chicks. At 4 of these nests, the adults ceased incubation of otherwise viable eggs when the first chick was 6 days old. Moreover, the most successful asynchronous nests tended to be those with the least hatching asynchrony, though the pattern was not significant.

Chicks that died, in both control and asynchronous broods, died at younger ages than chicks in synchronous broods (Fig. 2; median age: synchronous—day 12, control—day 7.5, asynchron-

TABLE 7
CHICKS RAISED TO 35 DAYS FROM BROODS WITH MANIPULATED SYNCHRONY

	Number of nests	Number of chicks fledged 0	1	2	3	Mean ± SD
Synchronous	32	3	6	14	9	1.91 ± .93
Control	31	3	5	13	10	1.97 ± .95
Asynchronous	20	2	4	5	9	2.05 ± 1.05[a]

[a] Artificially inflated (see text).

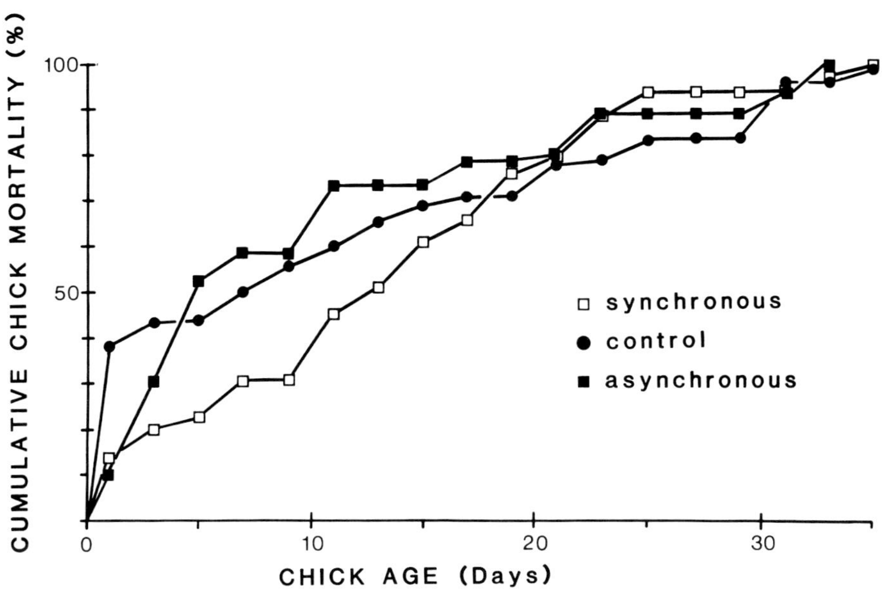

FIGURE 2. Cumulative percent mortality of chicks as a function of chick age. Sample size: synchronous, n = 35; control, n = 32; asynchronous, n = 19 chicks.

ous—day 5). The distribution of age at death differed significantly between synchronous and control broods (Kolmogorov-Smirnov, D = .248, n = 32, $P < .05$), and approached significance between synchronous and asynchronous broods (K-S, D = .298, n = 19, $P = .06$). In all groups, over 60 % of the chick mortality occurred by day 15.

The predicted advantage of a brood reduction strategy is that the early death of a chick that cannot be raised to fledging will result in more rapid growth of the remaining offspring. Thus, growth rates, in broods from which 2 chicks fledged, should be higher in control than in synchronous broods due to the earlier mortality among third-hatched control chicks. I compared the growth rates of chicks which subsequently fledged, among broods which fledged 2 chicks. Chick size and weight on day 20, but not on day 35, was significantly lower in synchronous broods (Table 8). There was a significant negative correlation between age of chick death and the size of surviving chicks on day 35 (Fig. 3).

In the context of the brood reduction hypothesis it is generally assumed that there is no advantage to asynchrony in broods where all 3 young can be raised; that is, the advantage should only be found in broods in which 2 chicks survive. In this experiment, however, the chick size and weight in nests from which all 3 chicks fledged, was significantly lower in synchronous nests than in controls on both day 20 and 35 (Table 9).

TABLE 8

GROWTH RATES OF CHICKS IN BROODS FROM WHICH TWO CHICKS FLEDGED[a]: MEAN ± SD (NUMBER OF NESTS)

	Age 20		Age 35	
Experimental group	Weight (g)	Signif. wt/size[b]	Weight (g)	Signif. wt/size
Synchronous	564 ± 140* (13)	.05/.03	895 ± 287 (3)	.15/.19
Control	637 ± 62 (13)		935 ± 84 (9)	
Asynchronous	571 ± 49* (4)	.03/.08	912 ± 67 (2)	.46/.39

[a] Raw data represents the average, for each nest, of the weight/size of chicks that survived to fledge. Weight/size was adjusted to the ages of 20 and 35. Differences from control nests are tested. All statistics were done on log transformed weights. Asymmetry resulting from the transformation of the standard deviation back to grams was averaged.

[b] One-tailed Mann-Whitney U.

* $P < .05$.

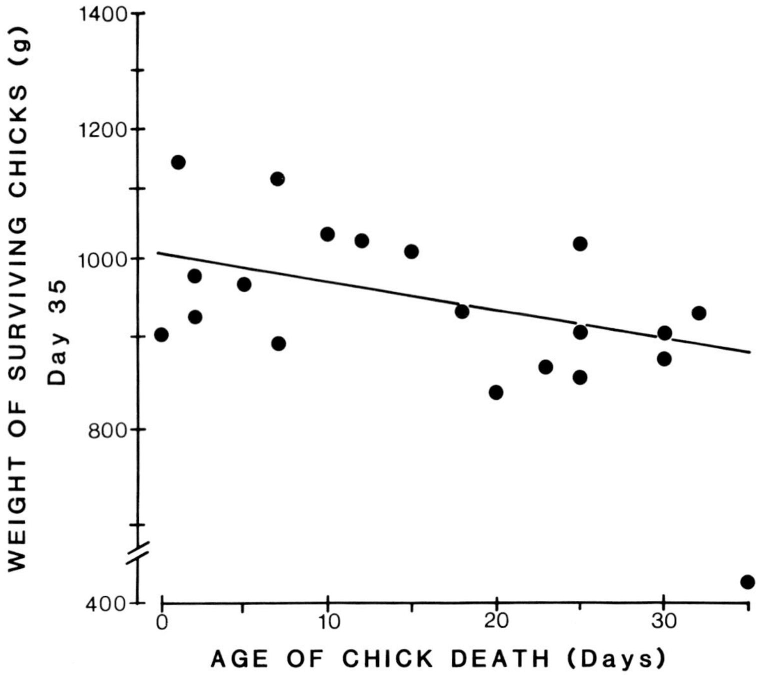

FIGURE 3. Average weight (g) at day 35 of surviving chicks (Log_{10} scale), in broods from which two chicks fledged, as a function of the age of death of the chick that died. Regression line excludes outlier. (Weight: Spearman rank r = −.39, *P* = .05, n = 19 excluding outlier; *P* = .02 including outlier; Size: r = −.42, *P* = .04, n = 19 excluding outlier; *P* = .01 including outlier).

DISCUSSION

Clutch-size Regulation

Two factors, incubation capacity and food limitation following the initiation of laying, could limit clutch size in the Glaucous-winged Gull. The shelf-life of eggs and the energetic cost of egg formation during the pre-laying period appeared to have little effect on clutch size.

Pre-laying energetics has received considerable attention as a factor influencing larid clutch size (Bateson and Plowright 1959, Lack 1968, Coulson and Horobin 1976, Nisbet 1977, Winkler 1983, 1985). Houston et al. (1983) have shown that the protein reserves of female Lesser Black-backed Gulls are correlated with both potential clutch size and egg weight. A connection between body condition and the timing of laying has been shown in the Ring-billed Gull (Boersma and Ryder 1983). Nisbet (1973, 1977) found that the amount of courtship feeding by Common Terns was correlated with subsequent total clutch weight and the weight of the c-egg and also found a correlation between female body weight at the

TABLE 9

Growth Rates of Chicks in Broods from Which Three Chicks Fledged[a]: Mean ± sd (Number of Nests)

Experimental group	Age 20		Age 35	
	Weight (g)	Signif. wt/size[b]	Weight (g)	Signif. wt/size
Synchronous	594 ± 124* (9)	.01/.01	910 ± 118* (3)	.02/.01
Control	708 ± 77 (9)		1037 ± 84 (8)	
Asynchronous	574 ± 60* (9)	.003/.005	931 ± 79* (8)	.03/.01

[a] Raw data represents the average, for each nest, of the weight/size of the three chicks. Weight/size was adjusted to the ages of 20 and 35. Differences from control nests are tested. All statistics were done on log-transformed weights. Asymmetry resulting from the transformation of the standard deviation back to grams was averaged.

[b] Two-tailed Mann-Whitney U.

* $P < .05$.

TABLE 10
NUMBER OF BROOD PATCHES AS RELATED TO CLUTCH SIZE FREQUENCIES[a]

Species	Clutch size frequency (%) 1	2	3	4	Ave. clutch	N	Brood patches	Reference
Fairy Tern *Gygis alba*	100				1.00	213	1	Dorward 1963
Swallow-tailed Gull *Creagrus furcatus*	100	*			1.00	>2000	2	Harris 1970
Royal Tern *Sterna maxima*	98.6	1.4			1.01	911	2	Buckley and Buckley 1972
Grey Gull *Larus modestus*	46.7	51.6	1.6		1.55	182	2	Howell et al. 1974
Ivory Gull *Pagophila eburnea*	31.2	68.7	*		1.69	32	3	Bateson and Plowright 1959
Kelp Gull	19	76	5		1.86	21	3	Williams et al. 1984
L. dominicanus	7.0	44.2	48.8		2.42	43	3	Unpub. data[b]
California Gull *L. californicus*	20.4	65.0	14.3	0.2	1.94	2032	3	Winkler 1983, Johnston 1956
Common Tern *S. hirundo*	7.7	23.4	68.0	0.8	2.62	951	3	Conover 1984[c], Gochfeld 1977
Glaucous-winged Gull *L. glaucescens*	2.7	21.2	76.1		2.73	704	3	This study
Black-headed Gull *L. ridibundus*	4.6	14.1	80.7	0.5	2.77	1217	3	Conover 1984[c], Beer 1961
Laughing Gull *L. atricilla*		18.0	82.0		2.82	111	3	Dinsmore and Schreiber 1974
Herring Gull *L. argentatus*	2.5	6.9	90.6		2.88	160	3	Drent 1970

[a] Data on clutch size is taken from same source as data on brood patch number where possible (same locality for California Gull).
[b] Punta Tombo, Argentina, November 1983.
[c] Post-1950 data only.
* Rare.

initiation of laying and clutch size. Energetic limitations may also provide an explanation for the reduced clutch size of the California Gull at Mono Lake (Winkler 1983, 1985).

There are several potential explanations for the absence of any apparent effect of supplemental food during the pre-laying period on egg size, laying date, or potential clutch size, in this study. First, the Glaucous-winged Gull is larger than other larids in which pre-laying energetics have been examined. The greater body size may serve to buffer the Glaucous-winged Gull from energetic factors immediately prior to laying. Second, food was provided for only 24 days prior to the median laying date of the a-egg and this may not have been sufficient time to have an effect. In other species (mostly passerines), significant advances in laying date have resulted when food was provided for 25–200 days prior to the mean laying date of controls (Ewald and Rohwer 1982). Finally, because of variation in the number of follicles that begin enlargement (Houston et al. 1983), increased energetic resources could result in the enlargement of a greater number of ova rather than increased allocation to each egg. In this case, however, the food-supplemented birds should have been more likely to lay a fourth egg; this was not observed. Spaans (cited in Drent and Daan 1980) reportedly found an advancement in laying date in Herring Gulls provided with supplemental food but there appear to be no other experimental data for the Laridae. Parsons (1976) argued that pre-laying energetics does not affect laying date in gulls based on the evidence that early nesting birds lay the largest eggs.

Supplemental food is known to affect breeding patterns in several other species of birds (Ewald and Rohwer 1982), although this is not always the case. Poole (1985) failed to find an effect of supplemental food on laying date or egg size in the Osprey (*Pandion haliaetus*) and Niebuhr (1981) found no correlation between courtship feeding and laying date in the Herring Gull. Food supplementation may not increase energetic resources available to the female but may instead substitute for courtship feeding and foraging. Food provided to incubating Herring Gulls results in an increase in time spent on territory (Shaffery et al. 1985). If benefit of increased egg size or earlier nesting is less than the benefit of territorial presence prior to laying, then increased energetic resources may be directed to the latter use.

Both this study and that of Parsons (1976) have found a correlation between laying date and the ability to produce a d-egg. This correlation is consistent with an energetic explanation if earlier breeders are in better condition (Boersma and Ryder 1983); however, this pattern could also result if, among late breeders, the advantage of earlier hatching exceeds the advantage of a third egg. Parsons (1976) found that birds that laid more than 3 eggs tended to lay larger first eggs. This pattern was not observed in this study, possibly because egg size does not appear to be correlated with laying date on Protection Island.

Incubation capacity may play a role in the regulation of clutch size in the Glaucous-winged Gull but its relative importance is questionable. More than 3 chicks can be hatched from artificially enlarged clutches, though hatching success per egg declines sharply among enlarged clutches. Because this study and that of Coulter (1973a, b) are indicative only of the proximate effect of incubation capacity on clutch size determination, its role as an ultimate constraint on clutch size is even more questionable.

There are 3 groups of birds in which incubation capacity has been a prominent hypothesis in the explanation of patterns of clutch size frequencies: the Charadrii (shorebirds), the Stercorariidae (jaegers and skuas) and the Laridae (Lack 1947, Klomp 1970, Andersson 1976, Winkler and Walters 1983). Some of the species in each group exhibit a truncated clutch size distribution, and yet have been shown to have the energetic resources necessary for production of extra eggs and the ability to fledge extra young. The incubation capacity hypothesis seems most plausible in the shorebirds, where the extremely large egg size, relative to body size, may place a physical limit on the number of eggs that can be incubated. Even here, though, experimental evidence for limits imposed by incubation capacity is not conclusive (Shipley 1984). In the Laridae and Stercorariidae the argument for incubation capacity as an ultimate limitation must rely on genetic, physiological, and developmental constraints since other species of similar size are capable of incubating larger clutches (Rohwer 1985, Fredrickson 1969).

There are 2 arguments against incubation capacity as an ultimate limitation to clutch size in the Laridae. First, over evolutionary time, the number of brood patches appears to be a plastic trait (Table 10). Species with small average clutch size have fewer brood patches. Since loss of a trait is easier than evolution of a novel trait, this evidence of plasticity is weakened if the primitive condition was to have 3 brood patches (e.g., Lack 1968). Second, there may be mechanisms other than the evolution of a 4th brood patch (3rd in the case of the Stercorariidae) that would allow efficient incubation of extra eggs. Increased heat transfer to the eggs, coupled with more frequent movement of eggs in the nest, is one such mechanism. Boersma and Ryder (1983) have documented variability in the vascularization of brood patches in incubating Ring-billed Gulls and this could potentially have a genetic basis. Enlargement of existing brood patches could also increase incubation efficiency. Drent (1970) found variability in brood patch size in incubating gulls; however, this may be attributed, at least in part, to differences in stage of incubation (F. Pitelka, pers. comm.). Because of the large egg size, a substantial increase in brood patch size would be necessary before 2 eggs could be incubated with a single brood patch.

In both this study and Coulter's (1973a, b), the benefit of more than 3 eggs, in terms of number of chicks hatched, was small. Thus, cost to adults of the production of a 4th egg would have to be small for a 4-egg clutch to result in a net benefit. In addition, I found that probability of loss of the entire clutch increased among enlarged clutches. Increased risk of a complete breeding failure may exceed the benefit of a 4th egg.

Mean hatching success may not be the most appropriate measure of the constraint imposed by incubation capacity. In 3 of the 20 4-egg clutches (15%) and 3 of the 18 5-egg clutches (17%), the entire clutch was successfully hatched. Existing variability in incubation behavior and physiology would seem to allow "good" pairs to lay extra eggs and successfully incubate them. Thus, incubation capacity does not impose a strict limit on clutch size, though it clearly decreases the marginal benefit of extra eggs.

In this study, the size of the c-egg was shown to be influenced by the amount of food available to the female following clutch initiation. There are currently 5 hypotheses that could account for the small size of the c-egg in gulls and terns. Three explanations assume that the reduced size of the c-egg is adaptive. Evidence for this assumption appears to be strong since the c-egg does not show a reduction in size following the removal of the a-egg, while the d-egg (if laid) is reduced in size (Paludan 1951, Parsons 1976). The female thus appears to have the energetic resources necessary to produce a large c-egg. First, the reduced size of the c-egg has been considered to be an adaptation for brood reduction (O'Connor 1978, Clark and Wilson 1981, Hahn 1981, Slagsvold et al. 1981). Clark and Wilson (1981) single out the small c-egg of gulls and terns as the only example of reduced egg size in which it appears that the reduction in size is an adaptation to impair the competitive ability of an offspring.

Second, Graves et al. (1984) argue that the c-egg represents an insurance egg only, hence the egg size is of little importance relative to energetic costs and time constraints. This explanation cannot suffice for all species which exhibit this trait since many populations frequently fledge three young (e.g., this study).

Finally, Parsons (1972, 1976) found that the length of incubation of the c-egg was relatively short and suggested that the small size functions to increase hatching synchrony, particularly when egg predation forces the early onset of incubation. There is no reason, however, to believe that gulls could not effectively guard eggs without transferring heat to them and thus it seems unlikely that predation rates could force asynchrony unless the probability of egg predation remained high even after the clutch was completed (Clark and Wilson 1981).

There have been 2 maladaptive explanations for the small size of the c-egg. First, Paludan (1951) suggested that the small size of the c-egg is simply a physiological response to the onset of incubation. Since other species of birds are capable of laying large last eggs after the onset of incubation (Clark and Wilson 1981), this argument has little support.

Second, Houston et al. (1983) have argued that the small c-egg is a result of a reduction in protein intake by the female after the onset of laying. A small c-egg has been found in every larid species investigated with the exception of the Black Tern (*Chlidonias niger*) and 1 population of the Western Gull (Pierotti and Bellrose 1986) (Table 11). The c-egg is characterized by the possession of the same amount of yolk but less albumen than the a- and b-eggs (Parsons 1976, Houston et al. 1983). Houston et al. (1983) found no correlation between female protein or lipid reserves and albumen weight and suggested that protein for albumen is derived from food intake between ovulation and laying (see Jones and Ward 1976). They suggest that the c-egg size is reduced because the female has less time available to forage after the onset of laying. Houston et al. (1983) cite the finding of a correlation between the extent of courtship feeding in Common Terns and the weight of the c-egg (Nisbet 1973) as evidence in support of this hypothesis. I calculated the weight of the c-egg relative to the average weight of the first 2 eggs from the data Nisbet (1973) presented and found no correlation ($r = .09$, $P > .25$, Spearman rank) between relative c-egg size and courtship feeding. The correlation between courtship feeding and third egg size was largely due to the strong correlation between total clutch weight and the size of the c-egg rather than an increase in relative c-egg size. While this is not inconsistent with the hypothesis of Houston et al. (1983) it provides little support. Pierotti and Bellrose (1986) found that under conditions of superabundant food, Western Gulls do not show a reduction in the size of the c-egg. This observation is consistent with the hypothesis of Houston et al. (1983) though, in the absence of an experimental study, it cannot be determined whether the large c-egg was a direct outcome of improved female condition or whether c-egg size was increased because the superabundant food indicated to the birds that there was no need for brood reduction adaptations.

The results of this study, showing an increase in the size of the c-egg following food supplementation on the day of clutch initiation, do provide support for the hypothesis of Houston et al. (1983). It appears that energetic resources are limiting the size of the c-egg. Thus, the small size may not be adaptive.

Since females are capable of laying large c-eggs if the first egg is removed, the most likely mechanism which could result in a protein shortage to the female is a change in foraging behavior as a result of the presence of an egg in the nest. In both gulls and terns, females tend to spend relatively more time on territory prior to laying than do males (Nisbet 1973, Pierotti 1981, Fitch and Shugart 1984, Maxson and Bernstein 1984). Courtship feeding during this period appears to be an important form of nutrition for the female (Nisbet 1973). An abrupt decline in courtship feeding at approximately the time of clutch initiation has been found in the Herring Gull and the Lesser Black-backed Gull (Brown 1967, Niebuhr 1981) though this does not appear to be the case in the Common Tern (Nisbet 1973). The reason for this reduction in courtship feeding is not clear. Fitch and Shugart (1984) found that both male and female attendance on territory increased during the fertile period (4 days prior to the laying of the a-egg up to the laying of the b-egg) and males spent more time with females during this period. These changes in attendance patterns could result in a decline in time available to both sexes for foraging, a decline in courtship feeding, and ultimately a decline in the size of the c-egg. The change in attendance patterns may be an adaptive mechanism by which the c-egg is made smaller; however, this would be an extremely circuitous mechanism since the female could just regulate food intake. The Black Tern, which does not show a reduction in c-egg size, exhibits relatively little nest guarding behavior between the initiation and completion of the clutch (Baggerman et al. 1956), and thus the female may be able to maintain a high protein intake. Similarly, the superabundant food available to the Western Gulls studied by Pierotti and Bellrose (1986) may provide the female with sufficient food despite changes in attendance patterns.

TABLE 11

RELATIVE EGG SIZE WITHIN CLUTCHES IN THE LARIDAE[a]

Egg volume index[b]	Second egg as % of first	Third egg as % of first	Number of clutches	Volume/ weight	Reference
Three-egg clutches:					
Ring Billed Gull (*L. delawarensis*)					
106.0	99.3[n]	96.7*	55	V	Ryder 1975[e]
103.4	100.2[t]	93.3[t]	43	V	Vermeer 1969
Common Gull (*L. canus*)					
97.4	100.0[n]	93.6*	138	V	Varth 1968
Herring Gull (*Larus argentatus*)					
152.7	97.4*	88.4*	453	V	Davis 1975
178.9	99.3[n]	90.3*	57	V[c]	Paludan 1951
159.3	99.1[n]	89.6*	50	V	Parsons 1975[b]
154.6	100.4[n]	88.9*	100	V[c]	Harris 1964
184.3	100.1[n]	92.7*	59	V	Barth 1968
174.0	100.5[n]	92.4*	76	V	Barth 1968
181.1	100.1[n]	94.8*	18	V	Barth 1968
Lesser Black-backed Gull (*L. fuscus*)					
139.7	99.5[n]	90.6*	62	V[c]	Paludan 1951
143.5	100.0[n]	94.3*	59	V[c]	Harris 1964
130.5	95.5[n]	88.2*	8	V	Barth 1968
136.3	97.1[n]	88.8*	44	V	Barth 1968
146.0	99.1[n]	92.7*	68	V	Barth 1968
California Gull (*L. californicus*)					
142.3	100.6[n]	96.5*	18	W	Behle and Goates 1957
137.6	98.4[t]	89.4[t]	55	V	Vermeer 1969
Western Gull (*L. occidentalis*)					
165.6	97.7[n]	90.7*	32	V	Coulter, this volume
182.0	100.0[n]	98.1[n]	103	W	Pierotti and Bellrose 1986
Kelp Gull (*L. dominicanus*)					
152.9	94.7[t]	86.6[t]	139	W[d]	Fordham 1964
167.3	97.1[n]	90.3*	8	W	Williams et al. 1984
166.2	98.4[n]	90.6*	18	V	Unpublished data[f]
Great Black-backed Gull (*L. marinus*)					
198.7	101.4[n]	98.0[t]	35	V[c]	Harris 1964
222.5	102.6*	97.1*	74	V	Barth 1968
211.7	99.9[n]	95.1*	93	V	Barth 1968
214.4	98.4[n]	95.2*	16	V	Barth 1968
Glaucous-winged Gull (*L. glaucescens*)					
174.2	98.8[n]	92.1*	152	V	This study
Laughing Gull (*L. atricilla*)					
75.7	97.7*	87.6*	137	W	Schreiber et al. 1979
82.7	97.1[n]	91.9*	15	V	Preston and Preston 1953
Silver Gull (*L. novaehollandiae*)					
70.6	95.8[n]	89.7*	12	V	Mills 1979
75.7	101.8[n]	92.8*	71	V	Wooller and Dunlop 1981[g]
Black-headed Gull (*L. ridibundus*)					
68.1	99.3[n]	94.3*	105	V[c]	Ytreberg 1956
70.5	99.9[n]	97.7*	156	V	Lundberg and Vaisannen 1979
Black-legged Kittiwake (*Rissa tridactyla*)					
88.7	97.3*	92.7*	33	V	Coulson 1963
92.0	99.4[n]	92.7*	26	W	Runde and Barrett 1981
Black Tern (*Chlidonias niger*)					
21.4	107.4[n]	102.7[n]	6	V	Dunn 1979

TABLE 11
CONTINUED

Egg volume index[b]	Second egg as % of first	Third egg as % of first	Number of clutches	Volume/weight	Reference
Common Tern (*S. hirundo*)					
38.4	96.3[t]	91.6*	112	V[c]	Gochfeld 1977
39.4	99.6[n]	97.9[n]	22	V	Gemperle and Preston 1955
n.d.	97.1[t]	94.6[t]	33	W[d]	Nisbet and Cohen 1975
n.d.	99.1[t]	96.7[t]	64	W[d]	Nisbet and Cohen 1975
Average (three egg):					
	99.1 ± 2.2	92.9 ± 3.5			
Two-egg clutches:					
Herring Gull (*L. argentatus*)					
154.3	94.3*		30	V	Parsons 1975b
Kelp Gull (*L. dominicanus*)					
161.4	98.0[n]		11	W	Williams et al. 1984
168.8	98.8[n]		14	V	Unpublished data[f]
Glaucous-winged Gull (*L. glaucescens*)					
174.7	96.6*		82	V	This study
Laughing Gull (*L. atricilla*)					
75.1	94.3*		71	W	Schrieber et al. 1979
Silver Gull (*L. novaehollandiae*)					
75.8	97.8[n]		51	V	Mills 1979
74.0	92.9*		238	V	Wooller and Dunlop 1981
Black-legged Kittiwake (*Rissa tridactyla*)					
87.8	95.9*		104	V	Coulson 1963
91.1	95.4[t]		77	V[d]	Maunder and Trelfall 1972
88.7	95.8*		67	W	Runde and Barrett 1981
93.2	96.2*		366	W	Runde and Barrett 1981
Roseate Tern (*S. dougallii*)					
n.d.	95.6[t]		63	W[d]	Nisbet and Cohen 1975
White-fronted Tern (*S. striata*)					
50.7	94.8*		34	V	Mills and Shaw 1980
Average (two egg):					
	95.9 ± 1.7				

[a] Significance as listed in source or calculated if possible (two-tailed t-test). If only length and breadth available, volume was calculated from mean values and listed as significant if both length and breadth were significant, not significant if neither was significant and not testable if one was significant. Data for more than one year are averaged (weighted) and listed as signifiacnt if any single year was significant. Data for different localities are listed separately.
[b] Volume (cc) is calculated as length × breadth;[2] volume is not corrected for shape and thus is an index only. Volumes are weighted averages of years for each egg, unweighted average of eggs.
[c] Variance for length and breadth only.
[d] No variance.
[e] Pairs with adult plumage only.
[f] Punta Tombo, Argentina, November 1983.
[g] Order of laying not certain.
* $P < .05$, different from a-egg.
[n] $P > .05$.
[t] Not testable.
n.d. No data.

The presence of a small b-egg in clutches of 2 eggs (see Table 11) does not refute this hypothesis as an explanation for the reduced c-egg size. A small b-egg could result from an earlier change in attendance patterns or from a shortage of energetic reserves in the female (Houston et al. 1983). Furthermore, the reduction in b-egg size, relative to the a-egg, is less in clutches of 2 (4.1%, n = 13 studies, 8 species, Table 11) than the reduction in c-egg size, relative to the a-egg, in clutches of three (7.1%, n = 40 studies, 15 species, Table 11).

This interpretation of the role of energy intake during laying provides one of the strongest mechanisms that could limit clutch size to 3 eggs. Females may not have a sufficiently high rate of protein intake following the initiation of the clutch to allow the laying of a fourth egg. Changes in attendance patterns that could increase the protein intake of the female may carry a cost for both adults in terms of an increased probability of egg predation, or a cost to the male resulting from a decreased probability of paternity.

The Brood Reduction Hypothesis

The results of this study indicate that the presence of brood reduction adaptations are compatible with the observation that clutch size is not limited by brood-rearing capacity. There was considerable variation in the extent to which the third chick was at a competitive disadvantage. Though in no case did all chicks hatch synchronously (n = 54), the difference in the time of hatching of the first and third chick was as little as 12 h (range 12–72 h). Pairs with relatively synchronous broods may have been those pairs capable of rearing enlarged broods. If the small c-egg is interpreted as a brood reduction adaptation (as the above analysis implies it should not be), the considerable variation in its relative size also is compatible with the view that some pairs were capable of raising enlarged broods. The volume of the c-egg was greater than or equal to the volume of the a-egg in 9.5% of clutches (n = 158, 1983–1985 combined). There was no correlation, however, between the relative size of eggs in a clutch and the degree of hatching synchrony, as could be expected if both traits are components of a brood reduction strategy.

Of greater importance, the results of this study indicate that asynchronous hatching may be beneficial regardless of the number of chicks raised. Among broods that fledged 2 chicks, the observed growth rates and timing of third chick mortality fit the predictions of the brood reduction hypothesis. There appeared to be no "cost" associated with asynchronous hatching, however, even among broods from which all chicks fledged. In contrast, synchronously hatched broods, from which all chicks fledged, had slower growth rates than control broods. The reduced growth rate in this situation could be considered further evidence in support of the brood reduction hypothesis (that is, only 2 chicks should have fledged in the slow growing synchronous broods but synchronous hatching prevented the early death of 1 chick) or else evidence suggesting that synchronous hatching would be a detriment even under conditions of abundant food.

In one other experimental study of asynchronous hatching in gulls a similar, somewhat paradoxical result was found. Hahn (1981) found that synchronous broods were less successful than control broods but not because of the predicted rapid mortality of third chicks in control broods; instead, in most control broods all chicks survived, while in most synchronous broods 1 chick died. Hahn (1981) also concludes that asynchronous broods may be more successful regardless of food supply.

SUMMARY

The truncation of the clutch-size frequency distribution of many members of the family Laridae traditionally has been considered to be a result of limits imposed by the incubation capacity of the adults. The incubation capacity of the Glaucous-winged Gull results in an abrupt decline in the marginal benefit of egg production after the third egg. Because of the asymmetrical benefit of a fourth egg relative to a third it is not surprising that the distribution of clutch sizes is skewed to the left. The incubation capacity hypothesis, however, cannot explain the virtual absence of 4-egg clutches since some pairs are capable of hatching up to 5 eggs. Moreover, incubation capacity represents only a proximate limit to clutch size since there is no *a priori* reason why brood patch number or size could not be increased to increase the hatching success of large clutches.

A more important limit to clutch size in gulls may involve energetic limitations following the initiation of the clutch. Females may not have a sufficiently high rate of protein intake following the initiation of the clutch to allow the laying of a fourth egg. This factor, particularly in conjunction with the decreased marginal benefit of egg production imposed by the incubation capacity, could explain the sharp truncation at 3 eggs and could be one of the selective factors leading to a modal clutch of 3.

I found no experimental evidence that clutch size is limited by energetic resources during the pre-laying period. The evidence from non-experimental studies, however, suggests that pre-laying energetics does influence egg production. The reason for this difference is not clear, but it would appear that energetic costs of egg production during the pre-laying period cannot be ruled out as another potential cost associated with the production of a fourth egg.

While this study found that the presence of brood reduction adaptations was compatible with the assumption that clutch size is not limited during the chick-rearing period, it also cast doubt on the general applicability of the brood reduction hypothesis to gulls and terns. The small c-egg, generally assumed to be an adaptation for brood reduction, may instead be a non-adaptive consequence of energy shortages during laying. Moreover, this study indicates that asynchrony may be advantageous regardless of the number of chicks that can be raised. Thus, asynchrony is apparently an adaptation for maximal growth under any circumstance rather than an adaptation for food stress.

ACKNOWLEDGMENTS

Special thanks are due to J. Lorenz, E. Cummins, and B. and P. Vorvick for providing solutions to innumerable logistical problems and for helping me obtain access to the island. Conversations with D. Boersma and S. Rohwer provided the impetus to address this question and led directly to at least one of the experiments; their comments greatly improved this paper. P. Caner, D. Stokes and S. Hiebert provided helpful comments on early drafts of the manuscript. I thank R. Morris, R. Schreiber, and K. Vermeer for their careful reviews of this paper. Funding was provided by the nongame program of the Washington Dept. of Game, Sigma Xi, the Chapman Fund, and National Science Foundation Dissertation grant BSR 84-00236.

LITERATURE CITED

AMLANER, C. J., JR., R. SIBLY, AND R. MCCLEERY. 1978. Effects of telemetry transmitter weight on breeding success in Herring Gulls. Biotelemetry 5:154–163.

ANDERSSON, M. 1976. Clutch size in the Long-tailed Skua *Stercorarius longicaudus*: some field experiments. Ibis 118:586–588.

BAGGERMAN, B., G. P. BAERENDS, H. S. HEIKENS, AND J. H. MOOK. 1956. Observations on the behavior of the Black Tern, *Chlidonias n. niger* (L.), in the breeding area. Ardea 44:1–70.

BARTH, E. K. 1968. Egg dimensions and laying dates of *Larus marinus, L. argentatus, L. fuscus,* and *L. canus.* Nytt. Mag. Zool. (Oslo) 15:5–34.

BATESON, P. P. G., AND R. C. PLOWRIGHT. 1959. The breeding biology of the Ivory Gull in Spitsbergen. Br. Birds 52:105–114.

BEER, C. G. 1961. Incubation and nest-building behavior of Black-headed Gulls. I. Incubation behavior in the incubation period. Behavior 18:62–106.

Behle, W. H., and W. A. Goates. 1957. Breeding biology of the California Gull. Condor 59:235–246.

BOERSMA, D., AND J. P. RYDER. 1983. Reproductive performance and body condition of earlier and later nesting Ring-billed Gulls. J. Field Ornithol. 54(4): 374–380.

BROWN, R. G. B. 1967. Courtship behavior in the Lesser Black-backed Gull, *Larus fuscus.* Behavior 29:122–153.

BUCKLEY, F. G., AND P. A. BUCKLEY. 1972. The breeding ecology of Royal Terns *Sterna* (*Thalasseus*) *maxima maxima.* Ibis 114:344–359.

CLARK, A. B., AND D. S. WILSON. 1981. Avian breeding adaptations: hatching asynchrony, brood reduction, and nest failure. Q. Rev. Biol. 56:253–277.

CONOVER, M. R. 1984. Occurrence of supernormal clutches in the Laridae. Wilson Bull. 96(2):249–267.

COULSON, J. C. 1963. Egg size and shape in the Kittiwake (*Rissa tridactyla*) and their use in estimating age composition of populations. Proc. Zool. Soc. Lond. 140:211–227.

COULSON, J. C., AND J. HOROBIN. 1976. The influence of age on the breeding biology and survival of the Arctic Tern *Sterna paradisaea.* J. Zool. (London) 178:247–260.

COULTER, M. C. 1973a. Clutch size in the Western Gull. Ibis 115:474.

COULTER, M. C. 1973b. Breeding biology of the Western Gull, *Larus occidentalis.* Unpubl. M.S. thesis, Oxford Univ. 76 p.

DAVIS, J. W. F. 1975. Age, egg-size and breeding success in the Herring Gull *Larus argentatus.* Ibis 117:460–473.

DINSMORE, J. J., AND R. W. SCHREIBER. 1974. Breeding and annual cyle of Laughing Gulls in Tampa Bay, Florida. Wilson Bull. 86:419–427.

DORWARD, D. F. 1963. The Fairy Tern *Gygis alba* on Ascension Island. Ibis 103b:365–378.

DRENT, R. H. 1970. Functional aspects of incubation in the Herring Gull. Pp. 1–132 *in* G. P. Baerends and R. H. Drent (eds.). The Herring Gull and its egg. Behavior Suppl. 17.

DRENT, R. H., AND S. DAAN. 1980. The prudent parent: energetic adjustments in avain breeding. Ardea 68:225–252.

DUNN, E. H. 1979. Nesting biology and development of young in Ontario Black Terns. Can. Field-Nat. 93(3):276–281.

EWALD, P. W., AND S. ROHWER. 1982. Effects of supplemental feeding on timing of breeding, clutch-size and polygyny in Red-winged Blackbirds *Agelaius phoeniceus.* J. Anim. Ecol. 51:429–450.

FITCH, M. A., AND G. W. SHUGART. 1984. Requirements for a mixed reproductive strategy in avian species. Am. Nat. 124(1):116–126.

FORDHAM, R. A. 1964. Breeding biology of the Southern Black-backed Gull. I. Pre-egg and egg stage. Notornis 11:3–34.

FREDRICKSON, L. H. 1969. An experimental study of clutch size of the American Coot. Auk 86:541–550.

GEMPERLE, M. E., AND F. W. PRESTON. 1955. Variation in the shape of eggs of the Common Tern in the clutch sequence. Auk 72:184–198.

GOCHFELD, M. 1977. Intraclutch egg variation: the uniqueness of the Common Tern's third egg. Bird-Banding 48:325–332.

GRAVES, J., A. WHITEN, AND P. HENZI. 1984. Why does the Herring Gull lay three eggs? Anim. Behav. 32:798–805.

HAHN, D. C. 1981. Asynchronous hatching in the Laughing Gull: cutting losses and reducing rivalry. Anim. Behav. 29:421–427.

HARRIS, M. P. 1964. Aspects of the breeding biology of the gulls *Larus argentatus, L. fuscus* and *L. marinus.* Ibis 106:432–456.

HARRIS, M. P. 1970. Breeding ecology of the Swallow-tailed Gull, *Creagrus furcatus.* Auk 87(2):215–243.

HARRIS, M. P., AND W. J. PLUMB. 1965. Experiments on the ability of Herring Gulls *Larus argentatus* and Lesser Black-backed Gulls *L. fuscus* to raise larger than normal broods. Ibis 107:256–257.

HAYMES, G. T., AND R. D. MORRIS. 1977. Brood size manipulations in Herring Gulls. Can. Zool. 55:1762–1766.

HOUSTON, D. C., P. J. JONES, AND R. M. SIBLY. 1983. The effect of female body condition on egg laying in Lesser Black-backed Gulls *Larus fuscus.* J. Zool. (Lond.) 200:509–520.

HOWELL, T. R., B. ARAYA, AND W. R. MILLIE. 1974. Breeding biology of the Gray Gull. *Larus modestus.* Univ. Calif. Publ. Zool. 104:1–57.

JOHNSTON, D. W. 1956. The annual reproductive cycle of the California Gull. II. Histology and female reproductive system. Condor 58:206–221.

JONES, P. J., AND P. WARD. 1976. The level of reserve protein as the proximate factor controlling the timing of breeding and clutch-size in the Red-billed Quelea *Quelea quelea.* Ibis 118:547–574.

KLOMP, H. 1970. The determination of clutch-size in birds, a review. Ardea 58:1–124.

LACK, D. 1947. The significance of clutch-size. I, II. Ibis 89:302–352.

LACK, D. 1968. Ecological adaptations for breeding in birds. Methuen. London.

LUNDBERG, C.-A., AND R. A. VAISANEN. 1979. Selective correlation of egg size with chick mortality in the Black-headed Gull (*Larus ridibundus*). Condor 81:146–156.

MAUNDER, J. E., AND W. THRELFALL. 1972. The breeding biology of the Black-legged Kittiwake in Newfoundland. Auk 89:789–816.

MAXSON, S. J., AND N. P. BERNSTEIN. 1984. Breeding season time budgets of the Southern Black-backed Gull in Antarctica. Condor 86:401–409.

MILLS, J. A. 1979. Factors affecting the egg size of Red-billed Gulls *Larus novaehollandiae scopulinus.* Ibis 121:53–67.

MILLS, J. A., AND P. W. SHAW. 1980. The influence of age on laying date, clutch size, and egg size of the White-fronted Tern, *Sterna striata.* N. Z. J. Zool. 7:147–153.

NIEBUHR, V. 1981. An investigation of courtship feeding in Herring Gulls *Larus argentatus.* Ibis 123: 218–223.

NISBET, I. C. T. 1973. Courtship-feeding, egg-size and breeding success in Common Terns. Nature 241: 141–142.

NISBET, I. C. T. 1977. Courtship-feeding and clutch-size in Common Terns *Sterna hirundo.* Pp. 101–109 *in* B. Stonehouse and C. Perrins (eds.). Evolutionary ecology (biology and environment vol. 2). Macmillan, London.

NISBET, I. C. T., AND M. E. COHEN. 1975. Asynchronous hatching in Common and Roseate Terns, *Sterna hirundo* and *S. dougallii.* Ibis 117:374–379.

O'CONNOR, R. 1978. Brood reduction in birds: selection for fratricide, infanticide and suicide? Anim. Behav. 26:79–96.

PALUDAN, K. 1951. Contribution to the breeding biology of *Larus argentatus* and *Larus fuscus.* Vidensk. Medd. Dan. Naturhist. Foren. 114:1–128.

PARSONS, J. 1970. Relationship between egg-size and post-hatching chick mortality in the Herring Gull (*Larus argentatus*). Nature 228:1221–1222.

PARSONS, J. 1972. Egg size, laying date and incubation period in the Herring Gull. Ibis 114:536–541.

PARSONS, J. 1975a. Asynchronous hatching and chick mortality in the Herring Gull *Larus argentatus.* Ibis 117:517–520.

PARSONS, J. 1975b. Seasonal variation in the breeding success of the Herring Gull: an experimental approach to pre-fledging success. J. Anim. Ecol. 44: 553–573.

PARSONS, J. 1976. Factors determining the number and size of eggs laid by the Herring Gull. Condor 78:481–492.

PIEROTTI, R. 1981. Male and female parental roles in the Western Gull under different environmental conditions. Auk 98:532–549.

PIEROTTI, R. 1982. Habitat selection and its effects on reproductive output in the Herring Gull in Newfoundland. Ecology 63(3):854–868.

PIEROTTI, R., AND C. A. BELLROSE. 1986. Proximate and ultimate causation of egg size and the "third-chick" disadvantage in the Western Gull. Auk 103: 401–407.

POOLE, A. 1985. Courtship feeding and Osprey reproduction. Auk 102:479–492.

PRESTON, F. W., AND E. J. PRESTON. 1953. Variation of the shapes of birds' eggs within a clutch. Ann. Carnegie Mus. 33:129–139.

ROHWER, F. C. 1985. The adaptive significance of clutch size in prairie ducks. Auk 102:354–361.

RUNDE, O. J., AND R. T. BARRETT. 1981. Variations in egg size and incubation period of the Kittiwake *Rissa tridactyla* in Norway. Ornis Scand. 12(1):80–86.

RYDER, J. P. 1975. Egg-laying, egg size and success in relation to immature-mature plumage of Ring-billed Gulls. Wilson Bull. 87:534–542.

SAMORODOV, YU. A., AND V. F. RYABOV. 1969. Kekologii chermogolovogo Khokhotung *Larus icthyaethus* Pull., kutanaiskoi oblasti. Vest. Mosk. Univ. Ser. 6 Biol. 24:32–37.

SCHREIBER, E. A., R. W. SCHREIBER, AND J. J. DINSMORE. 1979. Breeding biology of Laughing Gulls in Florida. Part I: Nesting, egg and incubation parameters. Bird-Banding 50:304–321.

SHAFFERY, J. P., N. J. BALL, AND C. J. AMLANER, JR. 1985. Manipulating daytime sleep in Herring Gulls (*Larus argentatus*). Anim. Behav. 33:566–572.

SHIPLEY, F. S. 1984. The 4-egg clutch limit in the Charadrii: an experiment with American Avocets. Southwest. Nat. 29:143–147.

SLAGSVOLD, T., J. SANDVIK, G. ROFSTAD, O. LORENTSEN, AND M. HUSBY. 1984. On the adaptive value of intraclutch egg-size variation in birds. Auk 101:685–697.

VERMEER, K. 1963. The breeding ecology of the Glaucous-winged Gull (*Larus glaucescens*) on Mandarte Island, B. C. Occas. Pap. B. C. Prov. Mus. 13:1–104.

VERMEER, K. 1969. Egg measurements of California and Ring-billed gull eggs at Miquelon Lake, Alberta, in 1965. Wilson Bull. 81:102–103.

WARD, J. G. 1973. Reproductive success, food supply and the evolution of clutch size in the Glaucous-winged Gull. Ph.D. thesis, University of British Columbia, Vancouver.

WEIDMANN, U. 1956. Observations and experiments on egg-laying in the Blackheaded Gull (*Larus ridibundus* L.). Brit. J. Anim. Behav. 4:150–161.

WILLIAMS, A. J., J. COOPER, AND P. A. R. HOCKEY. 1984. Aspects of the breeding biology of the Kelp Gull at Marion Island and in South Africa. Ostrich 55:147–157.

WINKLER, D. W. 1983. Ecological and behavioral determinants of clutch size: the California Gull (*La-*

rus californicus) in the Great Basin. Ph.D. thesis, University of California, Berkeley.

WINKLER, D. W. 1985. Factors determining a clutch size reduction in California Gulls (*Larus californicus*): a multi-hypothesis approach. Evolution 39:667–677.

WINKLER, D. W., AND J. R. WALTERS. 1983. The determination of clutch size in precocial birds. Curr. Ornithol. 1:33–68.

WOOLLER, R. D., AND J. N. DUNLOP. 1981. Annual variation in the clutch and egg sizes of Silver Gulls, *Larus novaehollandie*. Aust. Wildl. Res. 8(2):431–433.

YTREBERG, N. J. 1956. Contribution to the breeding biology of the Blackheaded Gull (*Larus ridibundus* L.) in Norway. Nest, eggs and incubation. Nytt. Mag. Zool. (Oslo) 4:5–120.

Studies in Avian Biology No. 10:26–43, 1987.

SEX RATIO SKEW AND BREEDING PATTERNS OF GULLS: DEMOGRAPHIC AND TOXICOLOGICAL CONSIDERATIONS

D. MICHAEL FRY,[1] C. KUEHLER TOONE,[1,2] STEVEN M. SPEICH,[3,4] AND R. JOHN PEARD[3]

ABSTRACT.—Female-female pairing and nests with supernormal clutches (SNC) of eggs occur in gull breeding colonies when two conditions are met: (1) sex ratio of breeding adults is skewed to females; and (2) nest sites are available for female-female pairs or polygynous trios. Expanding gull populations forming new colonies have SNC arising presumably from greater dispersion of females from their natal site. Declining populations in areas polluted with organochlorines (OC) show a decrease in breeding males and up to 15% of SNC. Population change has been predictive of both SNC incidence and sex ratio skew. SNC frequency and historical occurrence parallel high levels of OC pollution in southern California and the Great Lakes. OC exposure comparable to southern California has been experimentally duplicated by injecting OC into gull eggs, causing abnormal development of both male and female embryos. Males are feminized with germ cells located in the cortex of the gonad. Females develop both right and left oviducts. Persistent right oviducts, eggshell thinning and SNC in populations of gulls in Puget Sound, WA indicate recent exposure to estrogenic pollutants.

Female-female pairing in gulls was first described in southern California by Hunt and Hunt (1973, 1977) after examination of the sexes of birds attending supernormal clutches (SNC) of four to six eggs. Subsequent studies have documented SNC, female-female pairing, and polygynous trios among breeding populations of 4 species of gulls in North America: Western Gulls (*Larus occidentalis*) in small colonies on the Channel Islands in southern California; Herring Gulls (*Larus argentatus*) breeding in areas polluted with organochlorine pollutants in the Great Lakes; Ring-billed Gulls (*Larus delawarensis*) breeding in both polluted and unpolluted areas of the Great Lakes; and Ring-billed Gulls and California Gulls (*Larus californicus*) breeding in small colonies in Washington and Idaho (see review by Conover 1984a).

Ring-billed Gulls have been undergoing a rapid range and population expansion on islands in the Great Lakes (Ludwig 1974, Fox and Boersma 1983, Blokpoel and Tessier 1986) and in the Pacific Northwest (Conover et al. 1979b), and in both regions colonies have been described with female-female pairs (Fox and Boersma 1983, Ryder and Somppi 1979, Kovacs and Ryder 1981, 1983, Conover et al. 1979a). The analysis by Fox and Boersma (1983) indicates that female-female pairing in Ring-billed Gulls correlates with newly formed or rapidly expanding breeding colonies and that supernormal clutches are rare in stable colonies. The overall breeding population of California Gulls in the Pacific Northwest is similarly expanding (Conover et al. 1979b), and California Gull colonies with female-female pairing and SNC are similar to Ring-billed Gull colonies in that the birds move from year to year with changing conditions on islands in rivers.

The demographies of Western and Herring gull colonies with female-female pairs are markedly different from those of California and Ring-billed gulls. Colonies with SNC have been located in areas of high organochlorine contamination, and the breeding population of several of these colonies has decreased over the period from 1960–1975. Supernormal clutches of Western Gulls were first described by Schreiber (1970) and subsequently by Harper (1971), Hunt and Hunt (1973), Gress (1974), and Hand (1980). The Santa Barbara Island colony had an usually high incidence of supernormal clutches with as many as 14% of nests being attended by 2 females. The breeding population of Santa Barbara Island was severely impacted by organochlorine pollutants in the decade prior to the period of highest incidence of female-female pairing, and the breeding population of the colony declined from 3000 birds in 1972 to 850 in 1978 (Sowls et al. 1980).

Herring Gull colonies nesting in isolated areas of the Great Lakes also have had a high incidence of supernormal clutches and female-female pairing (Shugart 1980, Fitch and Shugart 1983). Reduced reproductive success of Herring Gulls occurred in colonies on Lake Michigan and Lake Ontario contaminated with organochlorine pollutants in the 1960s and early 1970s (Keith 1966, Hickey and Anderson 1968, Gilman et al. 1977, Mineau et al. 1984) and breeding success began to increase in parallel with reduced pollution during the late 1970s (Fitch and Shugart 1983,

[1] Department of Avian Sciences, University of California, Davis, California 95616.

[2] Present address: San Diego Zoo, Balboa Park, San Diego, California 92112.

[3] Cascadia Research Collective, Olympia, Washington 98501.

[4] Present address: P.O. Box 260-E, Malheur Field Stn, Princeton, Oregon 97721.

Mineau et al. 1984). The demographic parallels with Western Gulls are striking, as incidence of supernormal clutches or polygynous trios correlates with high organochlorine contamination, both geographically and temporally. Both species have had a marked increase in frequency of SNC since 1950, paralleling the use of organochlorine insecticides and the extent of organochlorine contamination in both southern California and the Great Lakes. Furthermore, stable breeding colonies of Western and Herring gulls in less polluted areas have not exhibited SNC or female-female pairing (Pierotti 1981, Burger and Gochfeld 1981, Nisbet and Drury 1984, Spear et al. this symposium).

The common factors in colonies of all 4 species of gulls with supernormal clutches appear to be the presence of a large surplus of breeding age females plus availability of suitable nest sites within the colonies. Nest site availability is covered in the discussion. Fox and Boersma (1983) and Fetterolf et al. (1984) postulated that new gull colonies formed during range expansion or colonization contain an excess of females due to greater dispersal of females from their natal sites. Male Herring Gulls are more philopatric than females (Chabrzyk and Coulson 1976) and thus, if other species of gull are similar, new colonies or colonies experiencing an influx of birds would be expected to have an excess of breeding age females (Greenwood 1980). Fox and Boersma (1983) concluded that SNCs result from behavioral responses to a skewed operational sex ratio accompanying range expansion and/or colonization.

Further evidence that female-female pairing is a result of a sex ratio skew is the experimental work of Conover and Hunt (1984a) who altered the sex ratio by trapping and removing adult male gulls from small Ring-billed and California gull colonies, thus creating an excess of females. The incidence of supernormal clutches and female-female pairing increased in these experimentally manipulated colonies, supporting the hypothesis that sex ratio skew encourages female-female pairing. The number of female-female pairs induced by artificial manipulation of the sex ratio was, however, considerably lower than the number of males removed, indicating that additional factors may play important roles such as nest-site availability (see discussion) and preference for male mates.

If sex ratio skew is a primary factor favoring the formation of female-female pairs and polygynous trios, an explanation for the cause of a large surplus of females in populations of Herring (Shugart 1980) and Western gulls (Hunt et al. 1980) must be found. The association of high levels of DDT in southern California, a decreasing breeding population of gulls, and a high proportion of female-female pairs were central to the hypothesis of Fry and Toone (1981) that organochlorines were involved with the abnormal sex ratios in these declining populations.

Gulls are moderately resistant to effects of DDT on eggshell thinning (Peakall 1975), and levels of total-DDT which caused severe eggshell thinning in pelicans and cormorants in southern California did not severely damage egg shells of gulls. Gull chicks in this region were hatched from eggs contaminated by relatively high concentrations of organochlorines. Keith (1966) described high levels of t-DDT in Herring Gull eggs in colonies on Lake Michigan where embryos survived concentrations as high as 200 ppm (fresh-egg weight basis). Direct embryo toxicity was observed at higher levels, but the reproductive tracts of exposed embryos were never examined.

While adult gulls are resistant to egg-shell thinning caused by organochlorines, gull embryos are quite sensitive to the teratogenic effects of estrogenic substances, which include several organochlorine pollutants and metabolites. The teratogenicity of estrogens to gulls was demonstrated by Boss (1943) and Boss and Witschi (1947) by injection of the synthetic estrogen, diethylstilbesterol (DES)(3,4-bis(*p*-hydroxyphenyl)-3-hexane), into Herring Gull eggs. Concentrations of 2.5 μg per egg (equivalent to 25 parts per billion in a 100 g egg), resulted in alteration of development and partial feminization of the reproductive tracts of males. The changes found in the reproductive tracts included development of ovarian cortical tissue in the testes, incorporation of primordial germ cells in the abnormal cortex of the gonad, and persistence of oviducts. Several estrogenic organochlorines have been identified including DDT and methoxychlor (Lutz-Ostertag and David 1973, review by Kupfer and Bulger 1980, Fry and Toone 1981), kepone (Eroschenko and Palmiter 1980, Eroschenko 1981), and mirex, which is metabolized to kepone (Eroschenko and Palmiter 1980). The estrogenic potential of DDT and high levels of DDT in gull eggs prompted Fry and Toone (1981) to speculate that the reproductive failures and sex ratio skew of breeding populations of Western Gulls could be a result of pollutant caused abnormal embryonic development.

We describe here results of injections of organochlorines (o,p'-DDT, p,p'-DDT, p,p'-DDE, and methoxychlor) into fertile eggs of Western and California gulls and the effects on survival and development of the reproductive system. In addition we present evidence that colonies of Glaucous-winged Gulls (*Larus glaucescens*) breeding in localized polluted areas of Puget Sound, WA, have exhibited egg shell thinning, persistent right oviducts in adult females, and a high frequency of supernormal clutches consistent with the hypothesis of organochlorine-induced abnormal development.

METHODS

Injection of Gull Eggs and Embryo Development

Collection of eggs and injection of organochlorines and estradiol

We obtained 317 eggs from two colonies: the Western Gull colony on Southeast Farallon Island (SEFI), CA, and the California Gull colony at Mono Lake, CA. Ninety-seven eggs were collected from SEFI in 1979, 110 collected from SEFI in 1980, and 110 eggs were collected from Mono Lake in 1980. One egg each from incomplete clutches were collected to maximize the probability that unincubated eggs were being collected. Eggs were packed in foam rubber in cardboard cases and transported to U.C. Davis within 12 hours of collection and stored at 10°C for 4 to 6 hours prior to injection and incubation.

Egg injections

Four organochlorines were injected into eggs. All were commercial compounds recrystalized by the Department of Environmental Toxicology, U.C. Davis, prior to use. Compounds injected were p,p′-DDT(1,1,-bis(4-chlorophenyl)-2,2,2-trichloroethane), Aldrich Chemical Co. "gold seal"; p,p′-DDE (2,2-bis(4-chlorophenyl) 1,1-dichloroethylene), Aldrich, 99%; o,p′-DDT (1-(2-chlorophenyl)1-(4-chlorophenyl)-2,2,2-trichloroethane), Aldrich "gold seal"; and methoxychlor (1,1,1-trichloro-2,2-bis(*p*-methoxyphenyl-ethane), City Chemical Corp. Compounds were injected at levels of 2, 5, 20, 50, and 100 μg/g egg wt. (fresh-egg weight basis), reflecting the range of pollutant concentrations found in pelican and gull eggs in southern California in 1970 and 1973. The estrogenic positive control, 17*B*-estradiol benzoate (1,3,5(10)-estratrien-3,17-diol 3-benzoate), Sigma Chemical Co., was injected in eggs at 0.5, 2, 5, and 20 ppm (fresh-egg weight basis). Compounds for injection were dissolved in corn oil and injected directly into the yolk. Total volume injected was 0.5% of total egg weight (92–104 g). Injection holes were covered with cellophane tape which prevented albumen leakage and entry of bacteria and fungi. Direct yolk injection was selected so as to mimic the distribution of fat-soluble pollutants in eggs. We successfully injected 264 eggs from which 108 embryos survived and developed to pipping. Control eggs were injected with corn oil alone. Yolks of 53 eggs were damaged during injection and excluded from the results.

Incubation program

Eggs were incubated in Marsh Farms incubators at 37°C and 50–55% relative humidity. Incubation period for California Gulls was 26 days; that for Western Gulls was 28 days. Chicks were removed from eggs and killed at hatching and examined for gross lesions. The reproductive tracts were photographed *in situ,* and gonads were fixed in buffered 10% formalin or 4% glutaraldehyde for histology. Tissues were embedded in paraffin or water-soluble plastic (Sorvall JB-4), sectioned, and stained with hematoxalin-eosin or basic fuschin-methylene blue. Serial sections of all left gonads were cut transverse to the long axis of the gonad, and each section was scored for the presence of primordial germ cells (PGC) in the cortex of the gonad. Testes and ovaries were sectioned and positive identification of testes of extensively feminized males was made from the presence of seminiferous tubules in the medullary portion of the gonad. Data from both species were pooled for analysis as no differences in the sensitivity of embryos or extent of developmental differences were observed between the 2 species. Fisher's Exact Test was used to determine the significance of differences between treatment groups (Zar 1984).

Survey of Glaucous-winged Gulls in Puget Sound

Glaucous-winged Gulls were studied in northern and western Washington as part of a 1984 National Oceanic and Atmospheric Administration (NOAA) survey of the population health of marine mammals and birds (Calambokidis et al. 1985). Gulls were collected from 5 breeding colonies: Goose Island in Gray's Harbor in western Washington; Smith Island in the Straits of Juan de Fuca 12 km west of Whidbey Island; Smith Cove at the north end of the Seattle waterfront; the St. Regis Paper Company, Tacoma; and the Simpson Timber Company, Shelton, at the southwestern end of Puget Sound. Two sites (Goose and Smith islands) were chosen as breeding colonies in relatively unpolluted areas and three sites (Seattle, Tacoma, and Shelton) were unusual colonies on roof tops and on piers in industrial areas and were selected on the basis of their historical or potential pollution impacts. The Seattle colony consists of several nesting locations on pilings, docks, and rooftops of warehouses and financial district offices (reviewed by Eddy 1982). The number of birds has grown slowly since 4 birds were first observed in 1936. Nearly 230 pairs nested on pier warehouses in 1981, but many of the warehouses were removed in 1982–1984 and the colony may therefore be nest-site limited at present (Eddy, pers. comm.). Surveys of the 5 breeding colonies were made during May, June, and July, 1984 (3–11 censuses per colony), during peak incubation.

Adult incubating females and their clutches of 3 eggs were collected during the first third of incubation. Adults were trapped on their nests and were weighed and measured to determine sex. In these colonies all birds weighing less than 1000 g were females, making separation easy. Females trapped on their nests were carried alive to the necropsy area. Blood was taken from the brachial vein for hematological studies, and the birds were killed with an inhalation overdose of ethyl ether. Gross necropsies were performed in the field, tissues were taken for residue analysis, and all remaining internal organs were fixed in 10% buffered formalin for histological examination. Complete gross and histopathological analyses were compiled for each bird from each study site and are presented elsewhere (Calambokidis et al. 1985).

All collected eggs were blown to remove contents which were retained frozen for residue analysis. Egg-

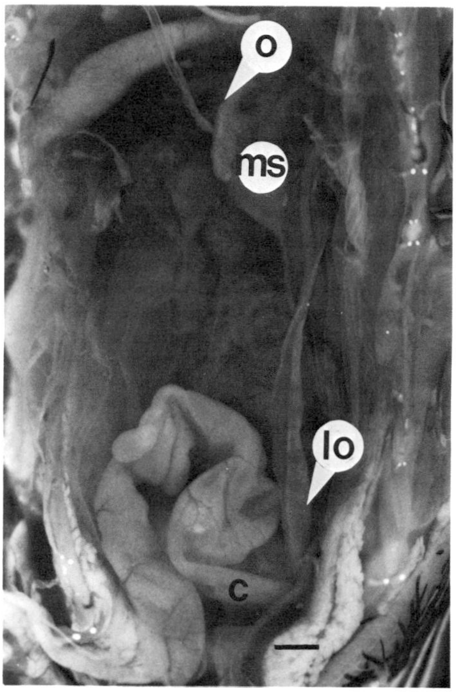

FIGURE 1. Reproductive organs of a partially eviscerated normal female Western Gull chick at hatching. A single left ovary (o) rests on the medial edge of the regressing mesonephros (ms). The left oviduct (lo) extends from the cloaca (c) to the border of the mesonephros. No right gonad is visible in this hatchling. (×3; scale bar 2 mm).

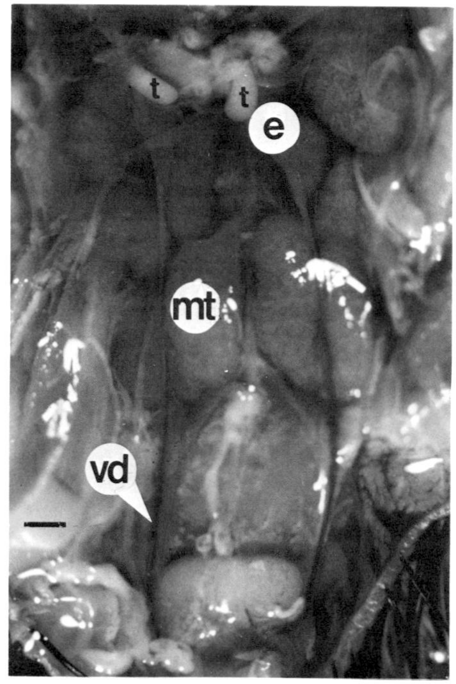

FIGURE 2. Reproductive organs of a control male California Gull chick at hatching. Left and right testes (t) are positioned at the cranial tip of the metanephros (mt). The mesonephros transforms into the epididymis (e) in males and the mesonephric ducts become the vasa deferentia (vd) which terminate at the cloaca. (×3; scale bar 2 mm).

shell thicknesses were measured by the Western Foundation for Vertebrate Zoology (WFVZ), Los Angeles, CA. Shell thickness was determined for each egg and compiled for each clutch. Percent egg-shell thinning was determined from comparisons with 52 Glaucous-winged Gull egg shells in the WFVZ collection which were gathered from the San Juan Islands, WA, prior to 1940.

RESULTS

Egg Injections and Embryo Development

Gross morphology of embryos at hatching

The anatomy of gull hatchlings and their response to estradiol or estrogenic compounds resembled those of chickens and Japanese Quail (Romanoff 1960, Lutz-Ostertag and David 1973), but gull embryos were more sensitive to the effects of estrogens, confirming the work of Boss (1943). Control female embryos had a prominent left ovary located at the cranial tip of the left kidney and a much reduced or absent right gonad (Fig. 1). Twenty of 21 control female embryos had a single left oviduct with a slightly thickened shell gland region adjacent to the cloaca. The left oviduct is formed by differentiation of the left Mullerian duct, with the right Mullerian duct normally regressing during embryogenesis. One control female embryo had a vestigial right oviduct represented as a 5 mm tubule attached on the right side of the cloaca.

The reproductive anatomy of control male embryos is shown in Figure 2. Paired testes of approximately equal size were located at the cranial tip of both left and right kidney. The embryonic mesonephros differentiates into the epididymis, and the mesonephric ducts become the vasa deferentia that empty into the cloaca. Both left and right Mullerian ducts regress in normal male embryos by the time of hatching. Testis position and size varied little in control embryos. Left and right testes were 0.8–1.2 mm in length and most testes were symmetrical in size. Some embryos had right testis approximately 30% smaller than the left testis.

Estradiol injections

Estradiol at the lowest concentrations injected (0.5 ppm) caused extensive feminization of em-

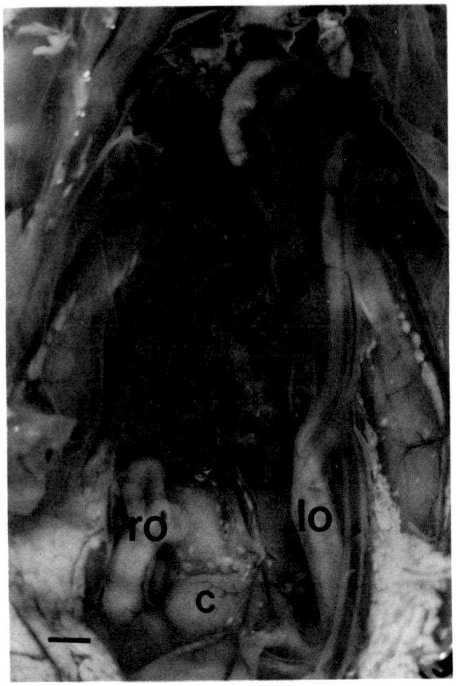

FIGURE 3. Female reproductive organs of a Western Gull chick injected with 0.5 ppm estradiol benzoate on day 1 of incubation. Both left oviduct (lo) and right oviduct (ro) extend from the cloaca (c). (×3; scale bar 2 mm).

bryos. The gross morphology of the reproductive tract exhibited both left and right oviducts in males and females (Fig. 3). The left oviduct was of normal female appearance, extending from the cloaca parallel to the left ureter and terminating in a funnel-shaped infundibulum adjacent to the left gonad. Right oviducts were less than half the length of left oviducts, frequently edematous, and sealed at the distal end. The gonads of both male and female embryos grossly resembled normal females, with 3–5 mm long left gonads and right gonads either not present or less than 1 mm in length. Males could not be separated from females by their gross morphology, and were identified from histological examination of sections of the gonad by the presence of seminiferous tubules in the left ovotestis (Fig. 11).

Organochlorine injections

The gross morphology of the reproductive tracts of embryos from eggs injected with organochlorine pollutants was intermediate between controls and estradiol injected eggs. O,p′-DDT (5 ppm and higher) and methoxychlor at high concentrations (20, 50, or 100 ppm) caused feminization of male embryos (Fig. 4) and persistence of right oviducts in surviving female embryos (Table 1). Feminization of male embryos was identifiable from the presence of left, or left and right oviducts, with the left oviduct extending from a modified left ovotestis to the cloaca. Left oviducts in feminized males usually had a thickened shell gland region adjacent to the cloaca (Fig. 4). Right oviducts were short, 2–6 mm long, and frequently edematous. Left gonads of feminized males were elongated, frequently showing a raised cortical ridge down the ventral midline of the ovotestis. Right testes of feminized males were either of normal size or reduced in size by as much as 50%. Low doses of o,p′-DDT and methoxychlor resulted in reduced abnormalities or no gross alteration of testes. The mixture of p,p′-DDE plus p,p-DDT (4:1) caused feminization of both male and females embryos at the high dose of 50 ppm (Fig. 5). Surviving embryos from eggs injected with p,p′-DDT or p,p′-DDE exhibited no gross abnormalities in either males or females.

Microanatomy of left gonads

In birds, males and female gonads develop from common primordia, but females differentiate in a sex-specific manner under the influence of estrogenic gonadal steroids. Without estrogenic hormonal influence, the primordial germ cells (PGC) migrate from the extra-embryonic membranes into the medullary tissue of the developing gonad which develops into a testis. Under the influence of estrogens, the PGC become localized in the cortex of the developing ovary. The histological anatomy of male and female gonads at hatching are distinct. Testes are circular or oval in cross section with a thin squamous epithelial cortex surrounding tortuous seminiferous tubules lined with prominently staining Sertoli cells and containing PGC that can be distinguished by their large nuclei and prominent nucleoli (Figs. 6, 7). PGC within seminiferous tubules remain in interphase and do not begin meiotic divisions into spermatogonia until the bird becomes an adult.

The left gonad of female embryos normally differentiates into a much flattened ovary under the influence of estrogen synthesized during differentiation (Fig. 8). Estrogens induce the PGC to become localized in a cortex of 100–150 μm thickness overlying medullary tissue consisting only of connective and vascular tissue. Primordial germ cells are prominent within the cortex and become organized in cords (Fig. 9). The morphology of the large primordial germ cells is distinctive with highly vacuolar cytoplasm and nuclei with condensed chromatin characteristic of meiotic prophase (Fig. 10). The large meiotic cells with condensed chromatin were used as a

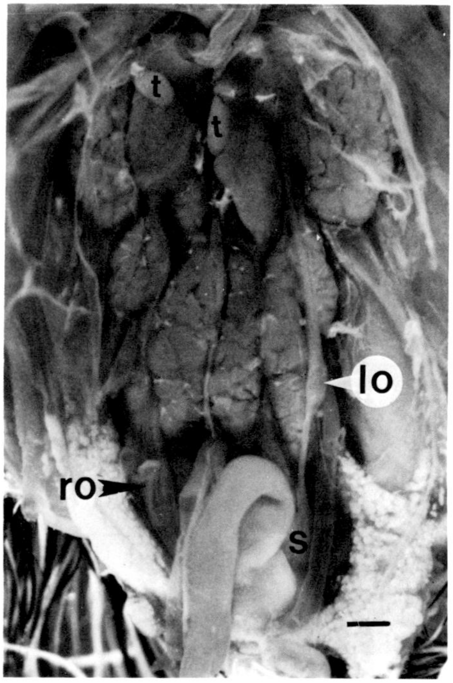

FIGURE 4. Reproductive tract of a male Western Gull hatchling injected with 5 ppm o,p′-DDT on day 1 of incubation. Testes (t) are of normal size and position. Feminization is indicated by presence of both short right oviduct (ro) and left oviduct (lo) with shell gland(s) at base of left oviduct (×3; scale bar 2 mm).

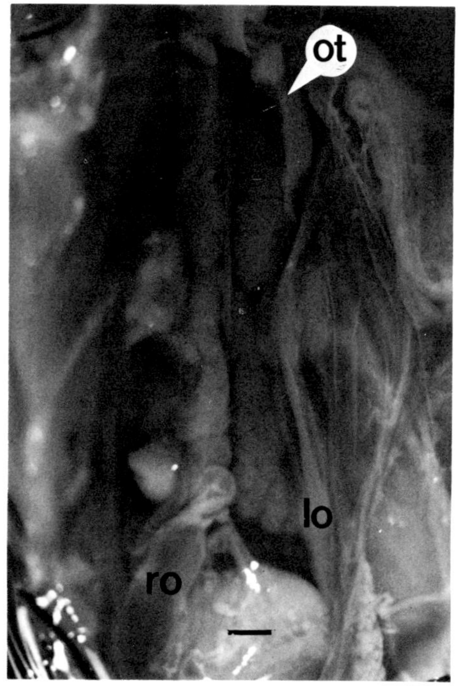

FIGURE 5. Reproductive tract of male Western Gull chick at hatching injected with a mixture of p,p′-DDT (10 ppm) and p,p′-DDE (40 ppm) on day 1 of incubation. The left gonad is enlarged and flattened into an ovotestis (ot). A left oviduct (lo) and an edematous short right oviduct (ro) are present. (×3; scale bar 2 mm).

marker of ovarian primordial germ cells. The PGC remain in prophase throughout juvenile and adult life and complete meiotic divison just prior to ovulation.

The gonads of both male and female embryos injected with estradiol grossly resembled ovaries. Male gonads, however, contained seminiferous tubules with occasional primordial germ cells that enabled positive identification of genetic sex (Fig. 11). The ovaries of female embryos injected with estradiol appeared normal.

The extent of feminization of male gonads of embryos exposed to organochlorines varied between compounds and was dose-dependent. The most sensitive indicator of feminization was localization of PGC in a thickened cortex on the surface of the left testis (Figs. 12 and 14). Primordial germ cells in the cortex of feminized testes entered meiosis and the condensed chromatin and vacuolar cytoplasm produced a positive feminization marker. We classified testes as feminized only when PGC arrested in meiotic prophase were present in the cortex of the testis. The thickness of the cortex was variable between individuals and between the two species and was not a reliable index of feminization. The lowest feminizing doses produced only scattered PGC in the cortex of the cranial tip of the testis, while increasing feminization resulted in clusters of PGC in a prominent ridge of cortical tissue extending the length of the elongated feminized left testis (Figs. 5, 12, 14). A second, but infrequently found, marker of feminization, described by Romanoff (1960), was the presence of seminiferous tubules which opened directly into the body cavity through the cortex of the testis (Fig. 13).

The most estrogenic DDT isomer was o,p′-DDT. Seven of 8 male embryos (87.5%) developed clusters of PGC in the cortex of the left testis at doses as low as 2 ppm (Table 1). Doses of 5 ppm and higher caused development of both left and right oviducts in addition to feminization of the cortex of the testes. The metabolite p,p′-DDE at high doses resulted in localization of primordial germ cells in the cortical tissue of half of the surviving male embryos. The mixture of p,p′-DDE and p,p′-DDT (4:1) at 50 ppm caused development of intersex testes. Methoxychlor induced the cortical localization of PGC

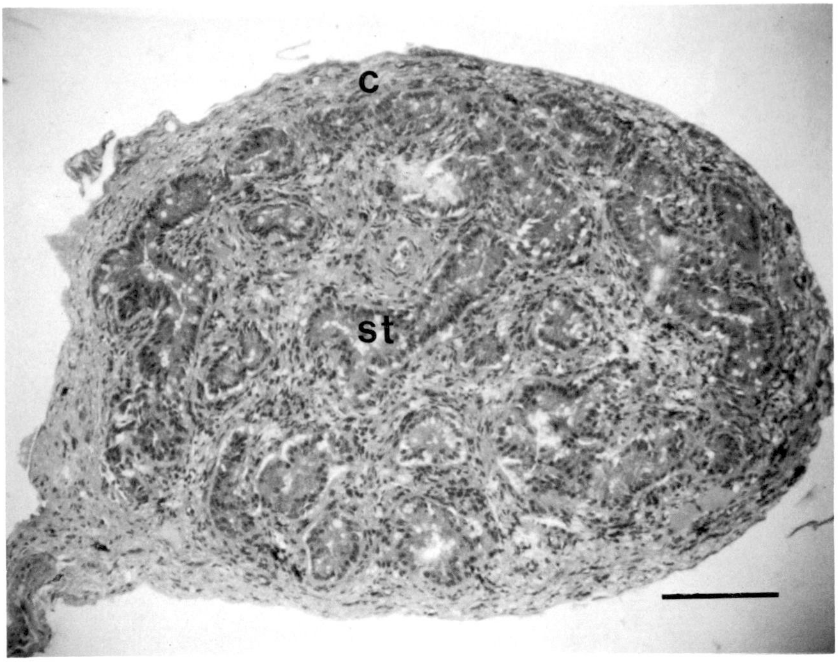

FIGURE 6. Left testis of a control male California Gull chick at hatching. The testis is oval in cross section with a thin fiberous cortex (c) surrounding a medulla containing many seminiferous tubules (st) cut obliquely and interspersed with interstitial connective tissue. (Paraffin embedded, hematoxalin and eosin stain (H&E); ×160, scale bar 100 μm.)

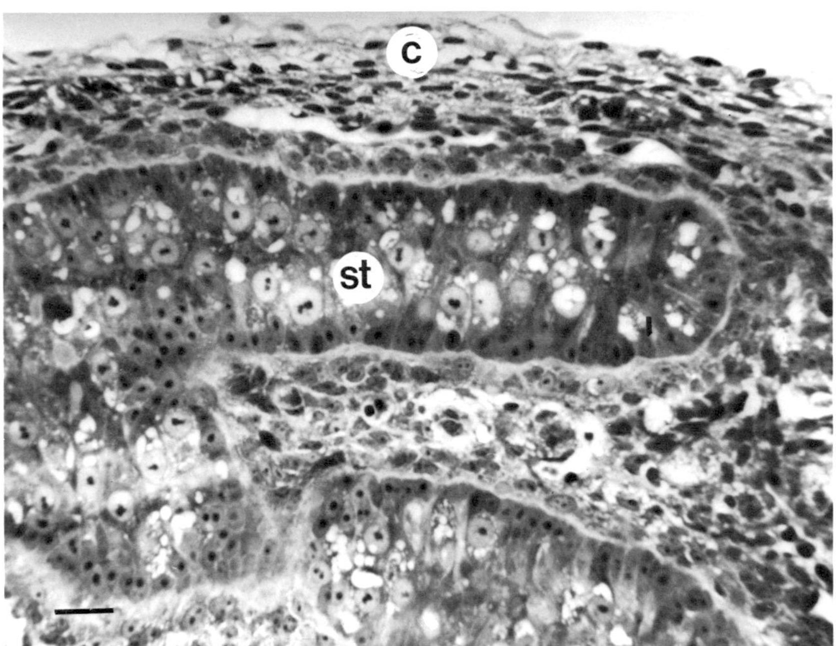

FIGURE 7. Left testis of a control male Western Gull chick at hatching. Seminiferous tubules (st) containing primordial germ cells with large interphase nuclei. Cortex (c) of testis is squamous epithelium. (Sorvall JB-4 embedded, basic fuschian and methylene blue stain (BF&MB); ×400, scale bar 20 μm.)

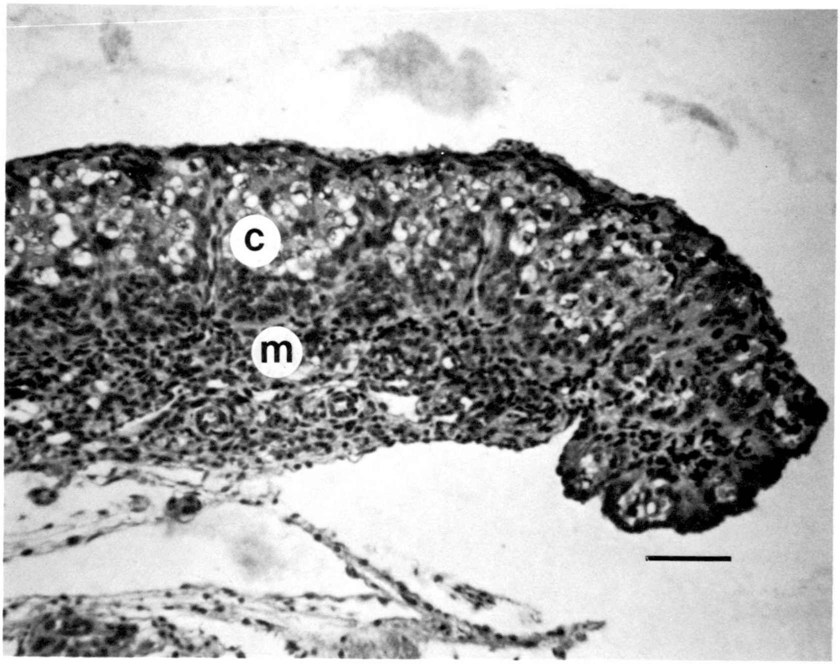

FIGURE 8. Left ovary of a control female California Gull chick at hatching. The ovary is flattened and differentiated into a cortex (c) containing primordial germ cells and a medullary region (m) containing vascular and connective tissue. (Paraffin embedded H&E; ×240, scale bar 50 μm.)

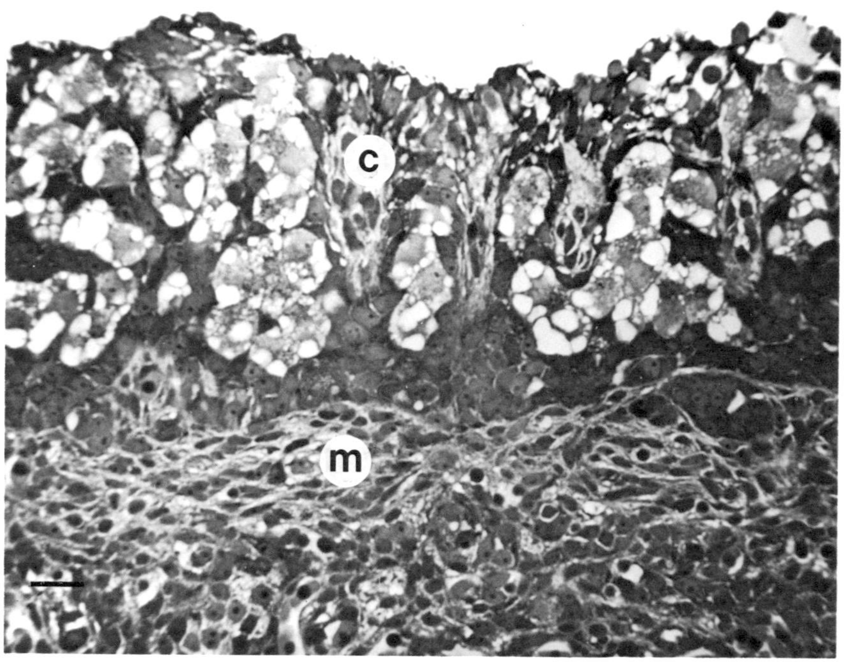

FIGURE 9. Left ovary of a control female Western Gull chick at hatching. Separation of cortex (c) and medulla (m) are distinct. Primordial germ cells are organized in cords within the cortex. (Sorvall JB-4, BF-MB; ×400, scale bar 20 μm.)

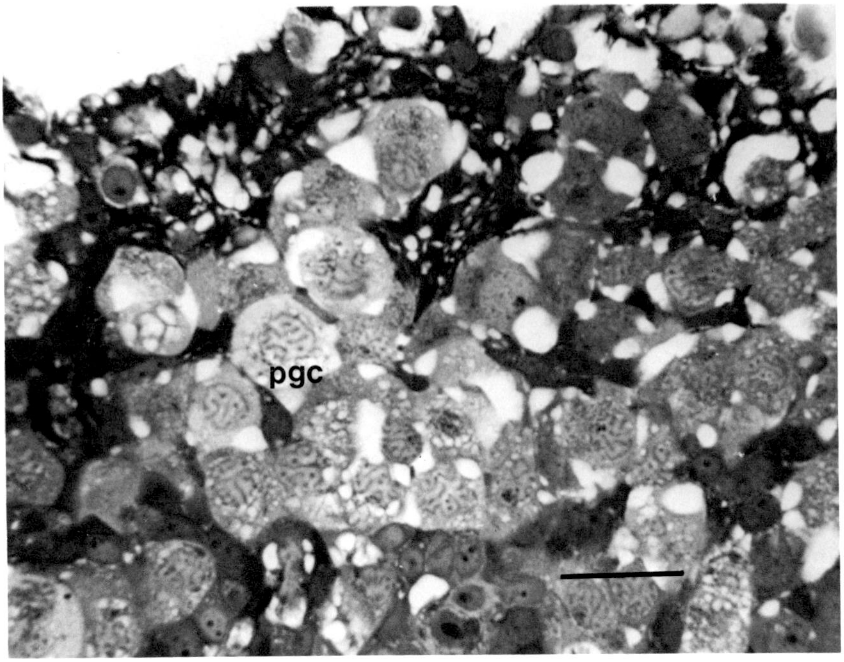

FIGURE 10. Cortex of left ovary of a control female Western Gull chick at hatching containing many primordial germ cells (pgc) with vacuolar cytoplasm and nuclei with thread-like condensed chromatin arrested in first meiotic prophase. (Sorvall JB-4, BF-MB; ×850, scale bar 20 μm.)

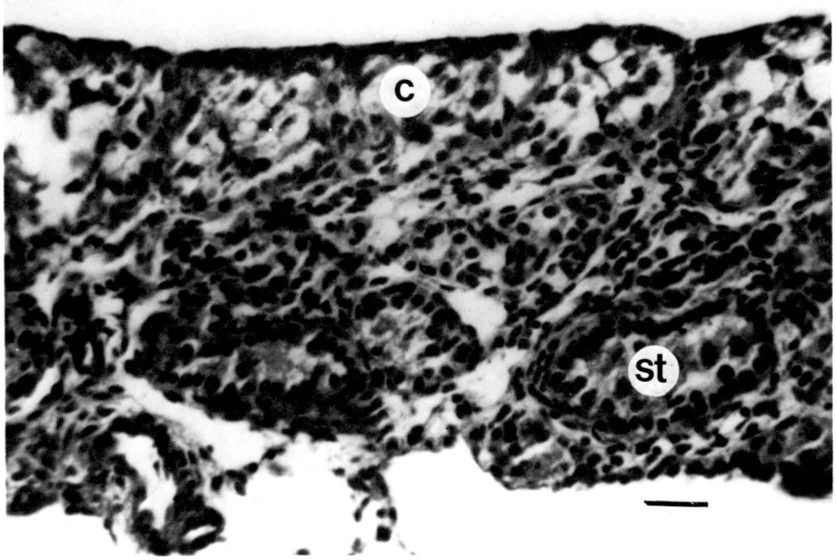

FIGURE 11. Left ovotestis of a male Western Gull chick at hatching injected with 2 ppm estradiol benzoate on day 1 of incubation. Testis has become flattened with a prominent cortex (c) containing primoridal germ cells with vacuolar cytoplasm. Seminiferous tubules (st) lined with Sertoli cells positively identify the organ as a feminized testis. (Paraffin embedded, H&E; ×440, scale bar 20 mm.)

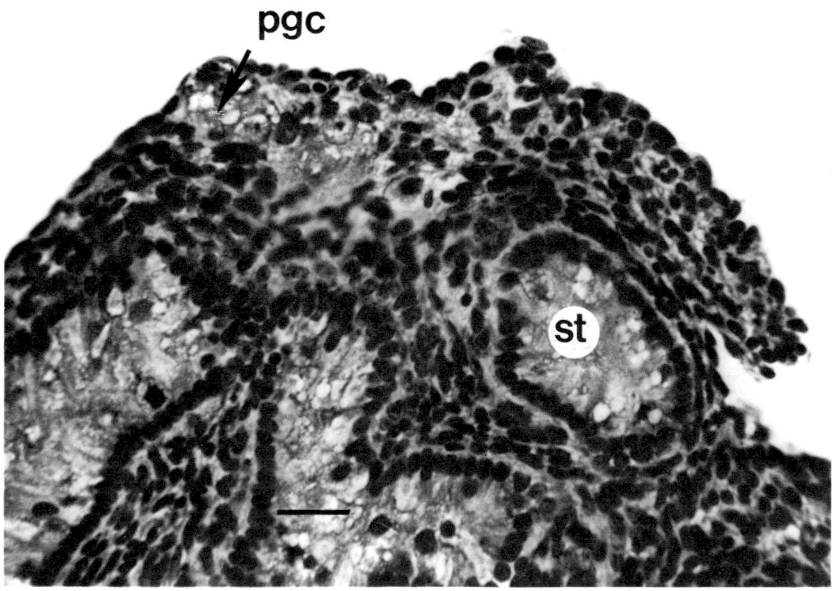

FIGURE 12. Left testis of a male Western Gull chick at hatching injected with 20 ppm o,p′-DDT on day 1 of incubation. Feminization is indicated by a thickened cortex containing clusters of primordial germ cells (pgc) arrested in first meiotic prophase. (Paraffin embedded, H&E; ×500, scale bar 20 μm.)

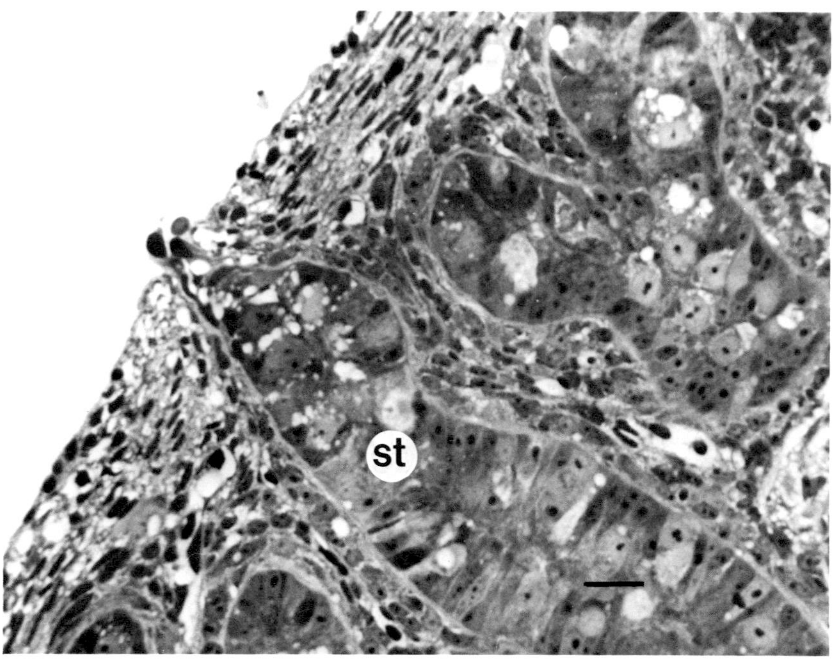

FIGURE 13. Left testis of a male Western Gull chick at hatching injected with 5 ppm o,p′-DDT on day 1 of incubation. A seminiferous tubule (st) opens through the cortex directly into the body cavity. (Sorvall JB-4 embedded, BF-MB; ×440, scale bar 20 μm.)

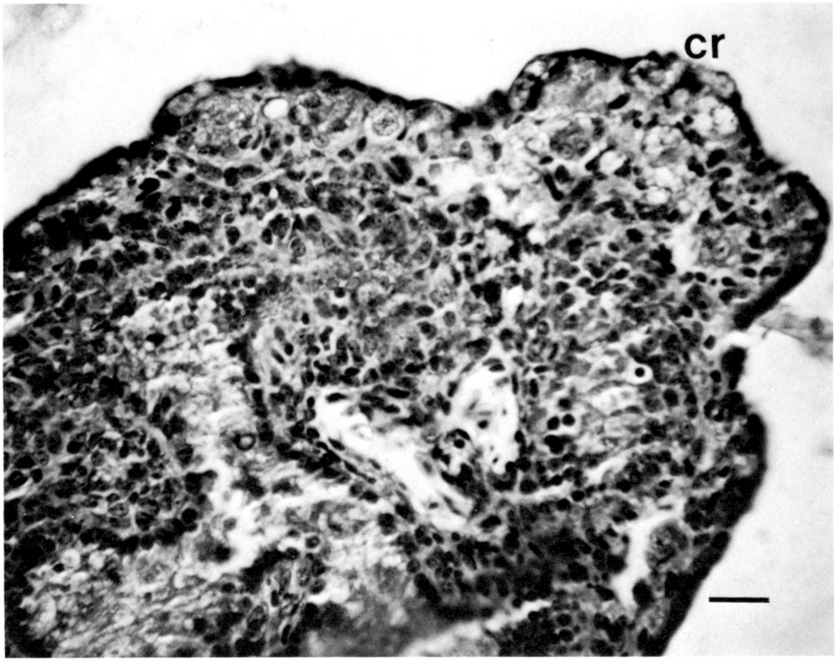

FIGURE 14. Left testis of a Western Gull chick at hatching injected with 100 ppm methoxychlor on day 1 of incubation. Feminization is indicated by thickening of the cortex into a cortical ridge (cr) containing many primordial germ cells arrested in first meiotic prophase. (Paraffin embedded, H&E; ×420, scale bar 20μm.)

at 2 ppm, and 4 of 5 male embryos surviving doses of 50 and 100 ppm developed both ovotestes and left and right oviducts (Fig. 14). p,p′-DDT did not result in feminization of any male embryos or persistence of right oviducts in any female embryos.

SURVEY OF GLAUCOUS-WINGED GULLS IN PUGET SOUND

Reproductive anatomy

The number of gulls collected in each area and reproductive tract comparisons are presented in Table 2. Female gulls trapped on nests with 3 eggs all had ovaries with 3 postovulatory follicle stalks and enlarged left oviducts that were beginning to regress. The extent of oviduct regression was correlated with the age of embryos observed when eggs were blown. Birds with the most regressed oviducts were incubating eggs with the largest embryos.

Twenty-one of 31 female gulls had right oviducts, 13 with right oviducts greater than 10 mm in length. The incidence and size of right oviducts observed in birds in this study varied with colony location. Most birds from Goose Island (5 of 6, 83%) and Seattle (5 of 6, 83%) had completely regressed or small right oviducts (less than 5 mm); this is the expected normal development. All birds from Tacoma (6 of 6) had right oviducts larger than 10 mm. Birds from Smith Island and Shelton were intermediate in both frequency and size of right oviducts.

Egg-shell thickness

Average egg-shell thicknesses of clutches combined by colony are given in Table 3. Egg-shell thicknesses from the two relatively unpolluted reference sites (Smith Island and Goose Island) were not significantly different from eggs collected in the San Juan Islands prior to the use of DDT (Table 3). Eggs from the 3 target sites of Seattle, Tacoma, and Shelton had average egg-shell thinning of 10%, 9%, and 8%, respectively. This is a marked amount of thinning for a gull species, and is comparable to thinning caused by high levels of DDT in Lake Michigan in the mid-1960s (Keith, 1966; Hickey and Anderson, 1968).

Supernormal clutches of eggs in Puget Sound

The mean clutch size in colonies of Glaucous-winged Gulls ranged from 2.21–2.86 eggs per nest (Table 4) and clutch size frequencies were similar to those recorded by Vermeer (1963) for Glaucous-winged Gulls breeding on Mandarte Island, B.C., in 1961. Three clutches of 4 eggs

TABLE 1
REPRODUCTIVE ANATOMY OF GULL EMBRYOS EXPOSED TO ORGANOCHLORINES OR ESTRADIOL BY INJECTION IN OVO

Treatment group	Dose	Normal males	Feminized males	Normal females	Abnormal females
Control		10	0	20	1
o,p'-DDT					
Low	2, 5	1	5**	2	2
High	20, 50, 100	0	2*	0	5***
p,p'-DDT					
Low	2	2	0	2	0
High	20, 100	2	0	3	0
p,p'-DDE					
Low	2	1	0	1	0
High	20, 100	3	3*	2	0
p,p'-DDT and p,p'-DDE (4:1)					
Low	5	1	1	2	1
High	50	0	2*	2	2
Methoxychlor					
Low	2, 5	0	3**	1	1
High	20, 50, 100	0	5***	1	6***
Estradiol					
Low	0.5, 2	0	5***	1	3**
High	5, 20	0	5***	1	0

* Different from controls $P < 0.05$.
** Different from controls $P < 0.01$.
*** Different from controls $P < 0.001$.
Significance determined using Fisher's Exact Test.
Data from Fry and Toone 1981.

were observed from a total of 1164 clutches. Two of the SNC were in Seattle (2.1% of clutches) and 1 at Goose Island (0.16% of clutches).

DISCUSSION

ESTROGENIC COMPOUNDS AND EMBRYONIC DEVELOPMENT

We have shown that estrogenic organochlorines are teratogenic and cause abnormal development of the reproductive tract in gulls. Exposure of domestic chicken (Wilson and Glick 1970), Japanese Quail (Adkins 1975, 1979), or Zebra Finch (Gurney and Konishi 1980) embryos to estrogens causes both feminization of the reproductive tract and "demasculinization" of the developing brain. The latter results in permanent suppression of sexual behavior in male birds when they become adults. Gulls are particularly susceptible to effects of estrogenic teratogens as shown by this study and that of Boss (1943) and the data support the conclusion that gulls are 10–50 times more sensitive than chickens, Japanese quail, or finches. It is not known whether the levels of estrogenic substances that cause gonadal changes in male gulls would also cause behavioral changes, but such changes occur in tandem in all other bird species which have been studied.

The estrogenic effects of DDT on developing chickens and quail parallel the results of this study. The greater sensitivity of gulls compared to chickens to the morphological feminizing effects of o,p'-DDT and methoxychlor has been confirmed in this study. Additionally, mixtures of p,p'-DDT plus p,p'-DDE also caused feminization, although requiring higher doses.

Exposure of embryos to estrogenic substances does not lead to complete sex reversal, but results in modification of the reproductive tract of both males and females. The breeding behavior of exposed females, however, is not altered by exogenous estrogens (see review by Ottinger et al. 1984). Female gull embryos in this study exhibited persistence of right oviducts with exposure to exogenous estrogens, and these changes may be associated with eggshell thinning and abnormal eggs when exposed birds become adults. Greenwood and Blyth (1938) found that chick-

TABLE 2
REPRODUCTIVE TRACT COMPARISONS OF GLAUCOUS-WINGED GULLS, WESTERN WASHINGTON, 1984[1]

Location	Number examined	Right oviducts present N	Right oviducts present %	Right oviducts >10 mm N	Right oviducts >10 mm %
Goose Island	6	1	17	1	17
Smith Island	6	5	83	3	50
Seattle	6	3	50	1	17
Shelton	7	6	86	2	29
Tacoma	6	6	100	6	100

[1] Data from Calambokidis et al. 1985.

TABLE 3
EGGSHELL THICKNESS (WITH MEMBRANE) OF WHOLE EGGS OF GLAUCOUS-WINGED GULLS, WESTERN WASHINGTON, 1984[1]

Location	Number eggs measured	Mean thickness mm (SD)	Per cent change from pre-1940[2]
Pre-1940, San Juan Islands	52	0.395 (.025)	—
Goose Island	33	0.388 (.029)	−2
Smith Island	16	0.384 (.035)	−3
Seatttle	13	0.354 (.032)	−10**
Shelton	29	0.362 (.025)	−8**
Tacoma	20	0.360 (.028)	−9**

[1] Data from Calambokidis et al. 1985.
[2] Measurements of eggs from the San Juan Islands.
** Significantly different from pre-1940 eggs from the San Juan Islands ($P < .001$).

TABLE 4
CLUTCH-SIZE DISTRIBUTION AND MEAN CLUTCH SIZE OBSERVED FOR NESTS OF GLAUCOUS-WINGED GULL COLONIES, WASHINGTON, 1984[1]

Location		Clutch size 1	2	3	4	Total clutches counted	Mean clutch size (±SD)	Survey dates
Goose Island	n =	133	219	259	1	612	2.21	8 June
	% =	22	36	42	<1		(0.78)	
Smith Island	n =	14	37	59	0	110	2.41	14 June
	% =	13	34	54	0		(0.71)	
Olympia	n =	0	3	27	0	30	2.90	May/June
	% =	0	10	90	0		(0.31)	
Seattle	n =	3	22	95	2	122	2.79	June/July
	% =	2	18	78	2		(0.50)	
Shelton	n =	8	29	83	0	120	2.63	18 June
	% =	7	24	69	0		(0.61)	
Tacoma	n =	4	21	175	0	200	2.86	June/July
	% =	2	11	88	0		(0.41)	

[1] Data from Calambokidis et al. 1985.

ens raised from estrone-injected eggs had abnormal oviducts and laid thin-shelled or abnormal eggs.

REPRODUCTIVE ABNORMALITIES AND CHEMICAL CONTAMINANTS IN PUGET SOUND

Both the persistence of right oviducts and eggshell thinning in Glaucous-winged Gulls in Puget Sound may be consequences of pollutant induced developmental abnormalities. Localized areas of Puget Sound have historical contamination, principally of PCB and polyaromatic hydrocarbons, but not extensive amounts of DDT (Riley et al. 1983). Inorganic and organic pollutants have been discharged into harbors of Puget Sound for many years (Malins et al. 1980, Riley et al. 1983). The major classes of contaminants in central and southern Puget Sound consisted of moderate concentrations of heavy metals (copper [Cu], zinc [Zn], arsenic [As], silver [Ag], antimony [Sb], mercury [Hg], and lead [Pb]), polychlorinated biphenyls (PCB), polynuclear aromatic hydrocarbons (PAH), and low to moderate concentrations of chlorinated butadienes. The highest concentrations of all pollutants were located in Commencement Bay adjacent to Tacoma and in Elliot Bay adjacent to Seattle. DDE was detected in most samples of fish and birds, but no DDT or DDD was detected in any samples. Riley et al. (1983) monitored contaminant levels in adult Great Blue Herons and in nestling Glaucous-winged Gulls, finding that Great Blue Herons from the Seattle and Tacoma areas had liver residues of PCB's of 1.0–5.1 ppm and concentrations in the fat of 14.6–80.4 ppm. Great Blue Herons from Sequim Bay on the straits of Juan de Fuca had much lower concentrations of PCB's (0.75 ppm liver, 5.5 ppm fat). Nestling gulls had low concentrations of PCB's, as would be expected from young birds, with levels in birds from Tacoma ranging from 0.81–1.1 ppm in liver and in Seattle ranging from 0.12–0.28 ppm in liver. Residues in gull nestlings from Protection Island were correspondingly lower (0.2–0.8 ppm). Small amounts of DDE were detected, but not quantified, in each of the samples of Great Blue Herons and gull nestlings.

The birds from urban areas of Puget Sound have had abnormally high concentrations of PCB relative to documented concentrations in marine birds from other areas of the world. The extent of organochlorine pollution in Puget Sound, as measured from tissue residues of herons and gulls in 1982, is low compared to organochlorine pollutants in parts of the Great Lakes and in southern California in the 1970s. Analysis of sediments in Puget Sound, however, indicates that historically higher levels of pollution existed prior to 1980 and that pollutants may have contributed to the death of marine birds through direct poisoning or through sublethal effects on birds at times of environmental stress (Riley et al. 1983).

The correlations between areas of moderate contamination, egg-shell thinning, and incidence of right oviducts is striking. In 1979 G. Eddy (pers. comm.) made a survey of the nests on warehouse roofs in Smith Cove, Seattle, and found 14 of 67 nests (21%) held clutches of 4 eggs. He repeated the survey in 1981 observing 147 nests, but none held 4 eggs. This extremely high incidence of SNC in 1979 remains unexplained. From the studies of Hunt et al. (1980), Kovacs and Ryder (1981, 1983), Fox and Boersma (1983), and Conover (1984b), the presence of SNC appears to accurately predict the occurrence of female-female pairing, but may underestimate the actual number of female-female pairs within a colony (Conover and Aylor 1985).

The incidence of SNC in Glaucous-winged Gulls in unpolluted areas is rare. Conover (1984a) compiled data from the literature and reported an occurrence of 0.1% for more than 5000 clutches. Vermeer (1963) working on Mandarte Island, B.C., observed zero SNC of 479 clutches in 1961. Galusha (pers. comm.) surveyed Glaucous-winged Gull nests on Protection Island in the Straits of Juan de Fuca and observed 1 SNC of 3909 clutches in 1980, 0 of 4057 in 1982, and 1 of 4734 in 1984.

The episode of extremely high density of supernormal clutches in 1979 and the continuing moderate number of supernormal clutches in Seattle is consistent with the distribution of pollutants. The actual causes of persistent right oviducts in females and egg-shell thinning in Puget Sound is unknown, however, as only trace amounts of DDE were detected by Riley et al. (1983) in surveys conducted in 1981 and 1982.

Organochlorine Pollutants and Gull Reproduction in Other Areas

The parallel distribution of organochlorine pollutants and female-female pairing correlates well with data from the Great Lakes and southern California, both in historical occurrence of SNC and geographical location. Incidence of SNC in Herring and Western gulls increased markedly subsequent to the introduction of DDT and other organochlorine insecticides after World War II (Conover 1984a, Conover and Hunt 1984b).

Polygyny and female-female associations of Herring Gulls occurred in northern Lake Michigan (Shugart 1980) in areas contaminated with high levels of DDT (Keith 1966, Strachan and Edwards 1984). The distribution of other Herring Gull colonies having polygyny and female-female pairing correlates geographically with the distribution of organochlorine pollution in Lake Ontario and the other Great Lakes (Gilman et al. 1977, Mineau et al. 1984, Gilbertson and Fox 1977, Fox et al. 1978, Fox and Boersma 1983).

Gulls and other seabirds in southern California were highly contaminated with DDT and DDE residues throughout the 1960s and early 1970s (Risebrough et al. 1967, Gress et al. 1973, Anderson et al. 1975). Southern California DDT pollution was particularly damaging because the estrogenic isomer o,p′-DDT was present in the food chain (MacGregor 1974) and was detected in Brown Pelican eggs (Lamont et al. 1970). MacGregor (1974) reported that o,p′-residues from myctophid fish averaged 22.3% of total-DDT throughout the period from 1949 to 1973 as DDT accumulated in southern California offshore waters. Total-DDT residues in fish averaged more than 4 ppm from 1966 through 1973. Total-DDT residues in Brown Pelican eggs in 1969 (Lamont et al. 1970) averaged about 80 ppm (wet-weight basis) with o,p-DDT and o,p′-DDD averaging 7.9% of the p,p′-isomers, a very significant amount of the estrogenic isomer. Brown Pelicans and Double-crested Cormorants showed extensive egg-shell thinning and reproductive failure due to organochlorine contaminants, but because gulls are moderately insensitive to egg-shell thinning by DDT (Hickey and Anderson 1968), continued breeding was possible and embryos were exposed to high levels of organochlorine contaminants. The overall effect of the organochlorine pollution in southern California was to cause the complete failure of reproduction in Brown Pelicans and cormorants and to reduce the productivity of the gull colonies.

The fitness and ultimate survival of gulls hatched from contaminated eggs in the wild is unknown. The demography of the gull colonies in southern California described by Hunt et al. (1980) indicates only a marked excess of breeding females. Without direct information from the field, it cannot be known whether male gulls hatched from contaminated eggs had decreased survival compared to females or whether feminization resulted in suppression of sexual behavior and self-exclusion from the breeding colony.

Reduced Numbers of Male Gulls in Breeding Populations

Sex-ratio skew in breeding populations of gulls in polluted areas has been postulated to be the result of selective mortality of males (Hunt et al. 1980, Shugart 1980, Nisbet and Drury 1983), although no evidence that male gulls are more sensitive to pollutants has been demonstrated. If male gulls, as adults, are more sensitive to organochlorine pollutants, a functional alteration in the sex ratio of the population could be achieved by differential survival of females in addition to the exclusion of developmentally feminized males from the breeding population. These possibilities are, of course, not mutually exclusive as both adults and embryos would typically be exposed to environmental contaminants and the reduced number of males in breeding colonies could reflect effects on embryos as well as on juveniles and adult males.

The presence of female-female pairs in a breeding colony suggests a skew in the operational sex ratio of the colony, but there are limited data on actual sex ratios in most colonies.

It has been suggested that sex ratios skewed to females in new or rapidly expanding colonies results because of a greater dispersion of females from natal colonies (Greenwood 1980, Fox and Boersma 1983, Fetterolf et al. 1984), although confirmation of the sex ratio has not been made in any of the Ring-billed Gull colonies studied. In the Western Gull colony on Santa Barbara Island which had a high incidence of female-female pairing and SNC, Hunt et al. (1980) determined the sex ratios of nesting pairs and non-breeding birds attending "bachelor clubs" and found a marked excess of females (19 females per male). Mills (1973), Burger and Gochfeld (1981), and Spear et al. (this symposium) have also documented surpluses of young breeding-age females in clubs of non-breeders, even in colonies that do not have supernormal clutches.

Nest Site Availability

As sex-ratio skew occurs in some colonies without female-female pairing, the additional factor of reduced competition for available nest sites appears to be of importance to allow the formation of female-female pairs and nesting. In areas of high breeding density, female gulls do not compete as successfully for territories as breeding males, Kovacs and Ryder (1981, 1983) reported that female-female pairs of Ring-billed Gulls breeding in high density colonies often have smaller territories than normal pairs and have greater difficulty in maintaining sites from year to year. Fetterolf et al. (1984) observed nest-site distribution within colonies and found female-female pairs in many colonies tended to be clustered, often occupying territories which became free during the season, such as spaces between large territories, or in areas of beach that became exposed after nesting had commenced.

Hand (1980) postulated that female-female pairs could establish territories even in relatively dense colonies as long as all breeding males also had adequate territories. In colonies with declining populations such as Santa Barbara Island, or in colonies with an artificially reduced number of males such as the colonies experimentally manipulated by Conover and Hunt (1984a), the availability of nest-sites is increased and affords female-female pairs a greater probability to establish breeding territories.

Even when a slight excess of breeding-age females exists, competition for territories may be responsible for the absence of female-female pairs in established, stable colonies of gulls. Female-female pairing and supernormal clutches are extremely rare in Atlantic maritime colonies (Nisbet and Drury 1984). A survey by Burger and Gochfeld (1981) in New England and New York, however, indicated that the composition of adult birds in established Herring Gull colonies averaged 52% females. "Bachelor clubs" of unpaired adults had a slightly greater sex-ratio skew of approximately 61% females. Similar data have been presented by Spear et al. (this symposium), who demonstrated a skewed sex ratio of non-breeding adult gulls on Southeast Farallon Island, CA, a very densely populated Western Gull colony with no occurrence of supernormal clutches. The sex ratio of the small number of non-breeders (about 700 out of a total population of 25,000) was 2.04 females per male. It is possible that given available nesting territory, some of the excess young females would form pairs and attempt to nest.

Why do surplus female gulls form female-female pairs? Both male and female gulls defend the breeding territories, although males are most vigorous in this activity (Pierotti 1981, Hand 1986). Normal female gulls have high circulating levels of androgens (Wingfield et al. 1980) that correlate with agressiveness and territoriality. This is not unusual, as females of most species that defend territories have levels of circulating androgens comparable to levels in males of the same species (Wingfield 1983). Females that enter into female-female pairs appear to be normal behaviorally as they have female behavior and do not adopt male behavior patterns in courtship or defense of territories (Hand 1981, Hunt et al. 1984). In colonies with a reduced number of males, female-female pairing by gulls appears to occur as a matter of course. It is indeed remarkable that at least some female gulls would persist in expending much energy when only rarely are female-female pairs successful in raising chicks (Kovacs and Ryder 1983, Hunt and Hunt 1977).

SUMMARY

Female-female pairing occurs in two demographic situations: (1) newly formed or rapidly expanding colonies of Ring-billed and California gulls nesting in the Great Lakes region and in the Pacific Northwest; and (2) populations of Herring, Western and Glaucous-winged gulls breeding in areas polluted with organochlorines. Influx into expanding colonies appears to occur with a sex ratio skewed to females because of increased dispersal of young breeding-age females from their natal sites. Sex-ratio skew of colonies in polluted areas may occur as a result of decreased fitness of male gulls hatched from contaminated eggs or decreased fitness of juvenile and adult male gulls exposed to organochlorine pollutants.

A skewed sex ratio alone is not sufficient to result in formation of female-female pairs. Reduced competition for available nest sites appears to be necessary if females are to defend territories successfully. Colonies having both of the required conditions (skewed sex ratio and available nest sites) are expanding colonies with excess nest sites and colonies with reduced populations in polluted areas where nest sites have become available due to a decrease in the male breeding population. Female-female pairing and reduced reproductive success of gulls in southern California and the Great Lakes is a striking example of pollutant effects on entire populations of birds.

ACKNOWLEDGMENTS

We thank R. Pierotti and D. Winkler for collecting eggs on S. E. Farallon I. and Mono Lake. We thank J. Calambokidis, J. Cubbage, and L. Lowenstine for assistance in the work on Glaucous-winged Gulls and G. Eddy for generously supplying data from Glaucous-winged Gull colonies in Seattle, WA. We thank L. Kiff and S. Sumida of the Western Foundation of Verebrate Zoology for measuring eggshells and providing data on shell thickness of Glaucous-winged Gull eggs from the San Juan Islands. This work was, in part, supported by NOAA Contract #83-ABC-00221.

LITERATURE CITED

ADKINS, E. K. 1975. Hormonal basis of sexual differentiation in the Japanese Quail. J. Comp. Physiol. Psychol. 89:61–71.

ADKINS, E. K. 1979. Effect of embryonic treatment with estradiol or testosterone on sexual differentiation of the quail brain. Critical period and dose-response relationships. Neuroendocrinology 29:178–185.

ANDERSON, D. W., J. R. JEHL, JR., R. W. RISEBROUGH, L. A. WOODS, JR., L. R. DEWEESE, AND W. G. EDGECOMB. 1975. Brown Pelicans: improved reproduction off the southern California coast. Science 190: 806–808.

BLOKPOEL, H., AND G. D. TESSIER. 1986. The Ring-billed Gull in Ontario: a review of a new problem species. Can. Wildl. Serv. Occ. Paper 67:1–32.

BOSS, W. R. 1943. Hormonal determination of adult characters and sex behavior in Herring Gulls (*Larus argentatus*). J. Exp. Zool. 94:181–209.

BOSS, W. R., AND E. WITSCHI. 1947. The permanent effects of early stilbestrol injections on the sex organs of the Herring Gull (*Larus argentatus*). J. Exp. Biol. 105:61–77.

BURGER, J., AND M. GOCHFELD. 1981. Unequal sex ratios and their consequences in Herring Gulls (*Larus argentatus*). Behav. Ecol. Sociobiol. 8:125–128.

CALAMBOKIDIS, J., S. M. SPEICH, J. PEARD, G. H. STEIGER, J. C. CUBBAGE, D. M. FRY, AND L. J. LOWENSTINE. 1985. Biology of Puget Sound marine mammals and marine birds: population health and evidence of pollution effects. NOAA Tech. Memorandum NOS OMA 18, p. 159.

CHABRZYK, G., AND J. C. COULSON. 1976. Survival and recruitment in the Herring Gull, *Larus argentatus*. J. Anim. Ecol. 45:187–203.

CONOVER, M. R., D. E. MILLER, AND G. L. HUNT, JR. 1979a. Female-female pairs and other unusual reproductive associations in Ring-billed and California gulls. Auk 96:6–9.

CONOVER, M. R., B. C. THOMPSON, R. E. FITZNER, AND D. E. MILLER. 1979b. Increasing populations of Ring-billed and California gulls in Washington state. West. Birds 10:31–36.

CONOVER, R. M. 1984a. Occurrence of supernormal clutches in the Laridae. Wilson Bull. 96(2):249–267.

CONOVER, M. R. 1984b. Frequency, spatial distribution and nest attendants of supernormal clutches in Ring-billed Gulls and California Gulls. Condor 86:467–471.

CONOVER, M. R., AND G. L. HUNT, JR. 1984a. Experimental evidence that female-female pairs in gulls result from a shortage of breeding males. Condor 86: 472–476.

CONOVER, M. R., AND G. L. HUNT, JR. 1984b. Female-female pairing and sex ratios in gulls: an historical perspective. Wilson Bull. 96(4):619–625.

CONOVER, M. R., AND D. E. AYLOR. 1985. A mathematical model to estimate the frequency of female-female or other multi-female associations in a population. J. Field Ornithol. 56(2):125–130.

EDDY, G. 1982. Glaucous-winged Gulls nesting on buildings in Seattle. Murrelet 63:27–29.

EROSCHENKO, V. P., AND R. D. PALMITER. 1980. Estrogenicity of kepone in birds and mammals. Pp. 305–325 *in* J. A. McLachlan (ed.). Estrogens in the environment. Proc. Symp. on Estrogens in the Environment, Raleigh, N.C., Sept.10–12, Elsevier/North-Holland, New York.

EROSCHENKO, V. P. 1981. Estrogenic activity of the insecticide chlordecone in the reproductive tract of birds and mammals. J. Toxicol. Environ. Health 8: 731–742.

FETTEROLF, P. M., H. BLOKPOEL, P. MINEAU, AND G. TESSIER. 1984. Incidence, clustering, and egg fertility of larger than normal clutches in Great Lakes Ring-billed Gulls. J. Field Ornithol. 55(1):81–88.

FITCH, M. A., AND G. W. SHUGART. 1983. Comparative biology and behavior of monogamous pairs and one male-two female trios of Herring Gulls. Behav. Ecol. Sociobiol. 14:1–7.

FOX, G. A., A. P. GILMAN, D. B. PEAKALL, AND F. W. ANUERKA. 1978. Behavioral abnormalities of nesting Lake Ontario Herring Gulls. J. Wildl. Manage. 42(3):477–483.

FOX, G. A., AND D. BOERSMA. 1983. Characteristics of supernormal Ring-billed Gull clutches and their attending adults. Wilson Bull. 95(4):552–559.

FRY, D. M., AND C. K. TOONE. 1981. DDT-induced feminization of gull embryos. Science 213:922–924.

GILBERTSON, M., AND G. A. FOX. 1977. Pollutant-associated embryonic mortality of Great Lakes Herring Gulls. Environ. Pollut. 12:211–216.

GILMAN, A. P., G. A. FOX, D. B. PEAKALL, S. M. TEEPLE, T. R. CARROLL, AND G. T. HAYMES. 1977. Reproductive parameters and egg contaminant levels of Great Lakes Herring Gulls. J. Wildl. Manage. 41(3): 458–468.

GREENWOOD, A. W., AND J. S. S. BLYTH. 1938. Experimental modification of the accessory sexual apparatus in the hen. J. Exp. Physiol. 28:61–69.

GREENWOOD, P. J. 1980. Mating systems, philopatry and dispersal in birds and mammals. Anim. Behav. 28:1140–1162.

GRESS, F., R. W. RISEBROUGH, D. W. ANDERSON, L.

F. KIFF, AND J. R. JEHL, JR. 1973. Reproductive failures of Double-crested Cormorants in southern California and Baja California. Wilson Bull. 85:197–208.

GRESS, F. 1974. Supernormal clutch sizes in Western Gulls breeding on the Channel Islands, CA. Report 32, Toxic Chemicals Section, Canadian Wildlife Service, Ottawa.

GURNEY, M. E., AND M. KONISHI. 1980. Hormone-induced sexual differentiation of brain and behavior in Zebra Finches. Science 208:1380–1383.

HAND, J. L. 1980. Nesting success of Western Gulls on Bird Rock, Santa Catalina Island, California. Pp. 467–473 *in* Dennis M. Power (ed.). The California Islands: Proceedings of a multidisciplinary symposium. Santa Barbara Museum of Natural History, Santa Barbara, Ca.

HAND, J. L. 1981. Sociobiological implications of unusual sexual behaviors of gulls: The genotype/behavioral phenotype problem. Eth. Sociobiol. 2:135–145.

HAND, J. L. 1986. Territory defense and associated vocalizations of Western Gulls. J. Field Ornithol. 57:1–15.

HARPER, C. A. 1971. Breeding biology of a small colony of Western Gulls (*Larus occidentalis wymani*) in California. Condor 73:337–341.

HICKEY, J. J., AND D. W. ANDERSON. 1968. Chlorinated hydrocarbons and eggshell changes in raptorial and fish-eating birds. Science 162:271–273.

HUNT, G. L., JR., AND M. W. HUNT. 1973. Clutch size, hatching success, and eggshell-thinning in Western Gulls. Condor 75:483–486.

HUNT, G. L., JR., M. W. HUNT. 1977. Female-female pairing in Western Gulls (*Larus occidentalis*) in southern California. Science 196:1466–1467.

HUNT, G. L., JR., J. C. WINGFIELD, A. NEWMAN, AND D. S. FARNER. 1980. Sex ratios of Western Gulls (*Larus occidentalis*) in southern California. Auk 97: 473–479.

HUNT, G. L., JR., A. L. NEWMAN, M. H. WARNER, J. C. WINGFIELD, AND J. KAIWI. 1984. Comparative behavior of male-female and female-female pairs among Western Gulls prior to egg-laying. Condor 86:157–162.

KEITH, J. A. 1966. Reproduction in a population of Herring Gulls (*Larus argentatus*) contaminated by DDT. J. Appl. Ecol. 3:57–70.

KOVACS, K. M., AND J. P. RYDER. 1981. Nest-site tenacity and mate fidelity in female-female pairs of Ring-billed Gulls. Auk 98:625–627.

KOVACS, K. M., AND J. P. RYDER. 1983. Reproductive performance of female-female pairs and polygynous trios of Ring-billed Gulls. Auk 100:658–669.

KUPFER, D., AND W. H. BULGER. 1980. Estrogenic properties of DDT and its analogs. Pp. 239–263 *in* J. A. McLachlan (ed.). Estrogens in the environment. Proc. Symp. Estrogens in the Environment, Sept. 10–12, 1979, Raleigh, N.C., Elsevier/North-Holland, New York.

LAMONT, T. G., G. E. BAGLEY, AND W. L. REICHEL. 1970. Residues of o,p'-DDD and o,p'-DDT in Brown Pelican eggs and Mallard Ducks. Bull. Environ. Contam. Toxicol. 5:231–236.

LUDWIG, J. P. 1974. Recent changes in the Ring-billed Gull population and biology of the Laurentian Great Lakes. Auk 91:575–594.

LUTZ-OSTERTAG, Y., AND D. DAVID. 1973. Action du DDT sur le tractus urogenital de l'embryon de Poulet et de Caille. C. R. Acad. Sc. Paris 276:1213–1216.

MACGREGOR, J. S. 1974. Changes in the amount and proportions of DDT and its metabolites, DDE and DDD, in the marine environment off southern California, 1949–72. Fishery Bull. 72(2):275–293.

MALINS, D. C., B. B. MCCAIN, D. W. BROWN, A. K. SPARKS, AND H. O. HODGINS. 1980. Chemical contaminants and biological abnormalities in central and southern Puget Sound. NOAA Tech. Memorandum OMPA-2, p. 295.

MILLS, J. A. 1973. The influence of age and pair-bond on the breeding biology of the Red-billed Gull, *Larus novaehollandiae scopulinus*. J. Anim. Ecol. 42: 147–162.

MINEAU, P., G. A. FOX, R. J. NORSTROM, D. V. WESELOH, D. J. HALLETT, AND J. A. ELLENTON. 1984. Using the Herring Gull to monitor levels and effects of organochlorine contamination in the Canadian Great Lakes. Pp. 425–452 *in* J. O. Nriagu and M. S. Simmons (eds.). Toxic contaminants in the Great Lakes. John Wiley & Sons, New York.

NISBET, I. C. T., AND W. H. DRURY. 1984. Supernormal clutches in Herring Gulls in New England. Condor 86:87–89.

OTTINGER, M. A., E. ADKINS-REGAN, J. BUNTIN, M. F. CHENG, T. DEVOOGD, C. HARDING, AND H. OPEL. 1984. Hormonal mediation of reproductive behavior. J. Exp. Zool. 232:605–616.

PEAKALL, D. B. 1975. Physiological effects of chlorinated hydrocarbons on avian species. Pp. 343–360 *in* R. Hague and V. H. Freed (eds.). Environmental dynamics of pesticides. Plenum Press, New York.

PIEROTTI, R. 1981. Male and female parental roles in the Western Gull under different environmental conditions. Auk 98:532–549.

RILEY, R. G., E. A. CRECELIUS, R. E. FITZNER, B. L. THOMAS, J. M. GURTISEN, AND N. S. BLOOM. 1983. Organic and inorganic toxicants in sediment and marine birds from Puget Sound. NOAA Tech. Memorandum NOS OMS 1, p. 123.

RISEBROUGH, R. W., D. B. MENZEL, D. J. MARTIN, AND H. S. OLCOTT. 1967. DDT residues in Pacific sea birds: a persistent insecticide in marine food chains. Nature (Lond.) 216:589–590.

ROMANOFF, A. J. 1960. The avian embryo. Macmillan, New York.

RYDER, J. P., AND P. L. SOMPPI. 1979. Female-female pairing in Ring-billed Gulls. Auk 96:1–5.

SCHREIBER, R. W. 1970. Breeding biology of Western Gulls (*Larus occidentalis*) on San Nicolas Island, California, 1968. Condor 72:133–140.

SHUGART, G. W. 1980. Frequency and distribution of polygyny in Great Lakes Herring Gulls in 1978. Condor 82:426–429.

SOWLS, A. L., A. R. DEGANGE, J. W. NELSON, AND G. S. LESTER. 1980. Catalog of California seabird colonies. Coastal Ecosystems Project, Fish and Wildlife Service, U.S. Department of the Interior, Washington, D.C., p. 371.

SPEAR, L. B., T. M. PENNIMAN, J. F. PENNIMAN, H. R. CARTER, AND D. G. AINLEY. 1987. Survivorship and mortality in a population of Western Gulls. Studies in Avian Biology 10:44–56.

STRACHAN, W. M. J, AND C. J. EDWARDS. 1984. Organic pollutants in Lake Ontario. Pp. 239–264 *in* J. O. Nriagu and M. S. Simmons (eds.). Toxic contaminants in the Great Lakes. John Wiley and Sons, New York.

VERMEER, K. 1963. The breeding ecology of the Glaucous-winged Gull (*Larus glaucescens*) on Mandarte Island, B.C. Pp. 1–104 *in* Occasional papers of the British Columbia Provincial Museum, No. 13. A. Sutton, Province of British Columbia.

WILSON, J. A., AND B. GLICK. 1970. Ontogeny of mating behavior in the chicken. Am. J. Physiol. 218: 951–955.

WINGFIELD, J. C., A. MARTIN, M. W. HUNT, G. L. HUNT, JR., AND D. S. FARNER. 1980. The origin of homosexual pairing of female Western Gulls (*Larus occidentalis wymani*) on Santa Barbara Island. Pp. 461–466 *in* D. M. Power (ed.). The California islands. Univ. California Press, Los Angeles.

WINGFIELD, J. C. 1983. Environmental and endocrine control of avian reproduction: an ecological approach. Pp. 265–288 *in* S. Mikami et al. (eds.). Avian endocrinology: environmental and ecological perspectives. Springer-Verlag, Berlin.

ZAR, J. H. 1984. Biostatistical Analysis. Second Edition. Prentice-Hall, Inc., Englewood Cliffs, New Jersey, 718 pp.

Studies in Avian Biology No. 10:44–56, 1987.

SURVIVORSHIP AND MORTALITY FACTORS IN A POPULATION OF WESTERN GULLS

LARRY B. SPEAR, TERESA M. PENNIMAN, JAY F. PENNIMAN, HARRY R. CARTER AND DAVID G. AINLEY[1]

ABSTRACT.—During 1978–1986, we studied survival and mortality in a large, stable (or slowly increasing) colony of Western Gulls (*Larus occidentalis*) breeding on Southeast Farallon Island, California. We monitored breeding adults returning to the colony, and examined 2073 dead birds of all ages at the colony and on the California coast from Monterey to the Oregon border. During the eight years, annual survival rates of adults four to 19 years of age averaged 84% for males and 81% for females. Estimated survival rates of first-, second-, and third-year gulls were 55%, 79%, and 85%, respectively. The sex ratio of young at dispersal was 0.92 M/F, however the sex ratio of fourth-year gulls was 0.68 M/F due to higher starvation rates of first-year males, and higher accident rates (amplified by factors related to food shortage) of second- and third-year males. A tendency for higher mortality rates among breeding females was probably related to competition with the larger males during periods of food shortage.

Knowledge of age- and sex-related survival rates is fundamental to understanding the dynamics of avian populations. Many estimates of survival rates in long-lived species are suspect, however, because of problems associated with band loss and use of samples provided by the public (reviewed by Botkin and Miller 1974, Coulson and Wooller 1976, Anderson et al. 1985). To our knowledge, only four studies provide reliable information on survival rates of adult gulls (Parsons 1971, Coulson and Wooller 1976, Chabryzk and Coulson 1976, Coulson and Butterfield 1986). We know of no estimates of survival rates of gulls in their first year that were not extrapolated from band recoveries, where it was necessary to assume survival rates of subadults (second- and third-year birds) as equal to that of adults (Chabryzk and Coulson 1976, and references therein). Finally, we have found no quantitative estimates of the different mortality factors that affect age or sex composition of gull populations.

The age and sex composition of Western Gull (*Larus occidentalis*) populations are of particular interest because of the apparent excess of adult females in breeding populations (Hunt et al. 1980, Pierotti 1981), and implications that this is a major factor leading to the phenomenon of female-female pairing in this and other larid species (reviewed by Conover and Hunt 1984, Hunt et al. 1984; see also Coulson and Thomas 1985). Sayce and Hunt (1987) hypothesized that the skewed sex ratio was primarily a result of higher mortality rates of males which, because of their larger size, may have more difficulty than females in obtaining enough food, particularly during the first months of independence. We report here the results of a study designed to examine a population of Western Gulls breeding on Southeast Farallon Island (SEFI), 42 km west of San Francisco, California, where female-female pairing has not been found (Pierotti 1981). Our objectives were to determine the sex-ratio at hatching, age- and sex-related survival rates, and mortality factors acting on age/sex groups.

STUDY AREA AND METHODS

COLONY STATUS

The number of Western Gulls breeding on SEFI was much reduced during the early 1900s, increased to an estimated 23,000 by 1959 (Ainley and Lewis 1974), and then to about 26,000 birds by 1986 (Point Reyes Bird Observatory—PRBO, unpubl.). The present population may be the largest ever, as the gulls nest densely over most of the 44 ha island (Coulter 1973, Pierotti 1981), including extensive areas formerly occupied by pinnipeds (Ainley and Lewis 1974). The colony represents approximatey 40% of the world's population of this species (reviewed in Spear et al. 1986).

Over 25,000 Western Gull chicks (approximately 2000 each year) have been banded on SEFI since DGA initiated a gull-banding program in 1971 (several thousand others were banded during the 1960s). It has become apparent that the SEFI population is essentially closed. Although 300–400 Western Gull young were banded at several other northern California colonies from 1975 to 1984, none are known to have immigrated to SEFI. Moreover, all Western Gull colonies within 600 km of SEFI have less than 150 pairs (Sowls et al. 1980), and only four banded Farallon gulls are known to have emigrated to these colonies since 1975 (K. Briggs, R. Lowe, and R. Pierotti, pers. comm.; Spear, pers. obs.).

SEX RATIO AT HATCHING

In 1984, we collected 99 young (one from each nest) immediately after hatching. These included 33 young each from eggs laid first, second, and third in the clutch. All were sexed by gonadal inspection.

SURVIVAL RATES OF FIRST-, SECOND-, AND THIRD-YEAR BIRDS (PREBREEDERS)

Between July 1978 and February 1986, LBS made 41 censuses during which 919 dead Western Gulls were

[1] Point Reyes Bird Observatory, Stinson Beach, California 94970.

found along the Pacific coast from Monterey, California to the Oregon border. Censuses for dead birds were made during January (6), February (5), March (3), April (1), May (1), July (4), August (4), September (4), October (3), November (6), and December (4). No census was made in June. We define the seasons as: summer (Jun–Aug), fall (Sep–Nov), winter (Dec–Feb), and spring (Mar–May). Search effort was standardized by location; each census included all coastal dumps and fishing ports, and about 300 km of beach traversed by foot or automobile. All 919 gulls were fresh enough to be examined as described below (see Methods: Mortality Factors on the California Coast). For dead gulls found on the coast, we considered age classes including adult, third-, second-, and first-year birds. We distinguished age classes by plumage differences (see McCaskie 1983) and arbitrarily set hatching dates at 1 July (see Coulter 1973 for breeding chronology).

Mathematical relationships for estimating survival rates of prebreeders have not been derived for use with data such as ours (see above). Therefore, to estimate the annual survival rates (SR) of first-year gulls ($SR^{0\text{-}1}$) we derived the relationship

$$1 - \left[\frac{(M_a)(N_1)}{(N_a)(P)}\right],$$

where M_a is the annual mortality rate of adults breeding on SEFI, N_1 is the number of first-year gulls in the coastal sample, N_a is the estimated number of adults of breeding status in the coastal sample, and P is the annual productivity rate (no. of young dispersed) of a breeding Farallon adult. Because some adults did not breed, the total coastal sample of adults included both breeding and nonbreeding birds. Therefore, to determine N_a it was necessary to exclude from the coastal sample the percentage of adults that were nonbreeders (see Methods: Number and Sex Ratio of Nonbreeding Adults, and Results: Population Status). We used values of N_a and M_a from adult males only because the population of nonbreeding adult males was more easily estimated than was that of nonbreeding adult females. The relationship above is similar to that derived by Coulson and White (1959), but eliminates possible bias resulting from the assumption that mortality rates of subadults and adults are the same. Survival rates of second-year ($SR^{1\text{-}2}$) and third-year ($SR^{2\text{-}3}$) birds were estimated from the relationships

$$1 - \left(\frac{N_2}{R_2}\right), \quad \text{and} \quad 1 - \left(\frac{N_3}{R_3}\right), \text{ respectively,}$$

where N_2 and N_3 are the respective numbers of second- and third-year birds in the coastal samples, R_2 is the estimated number of gulls surviving their first year and therefore at risk during the second, and R_3 is the estimated number of gulls surviving their second year and therefore at risk during the third. R values were calculated as follows: $R_1 = N_1/1 - SR^{0\text{-}1}$, $R_2 = R_1 - N_1$, and $R_3 = R_2 - N_2$.

Because movement patterns of Western Gulls originating from Oregon and Washington populations vary between different age classes (Coulter 1975), we adjusted the coastal sample of dead gulls to exclude these birds. The number of "out-of-state" birds in the study area was estimated from population censuses (Pitman et al. in press, Speich and Wahl in press) together with an analysis of these gulls' movements (Coulter 1975). Assuming that their survival rates were similar to those of resident Western Gulls, this information indicates that Oregon and Washington birds constituted $<1.0\%$, 6.0%, 6.0%, and 18.0%, respectively, of adult, third-, second-, and first-year gulls in our coastal sample. We also assumed that sedentary Western Gulls from southern California colonies (Coulter 1975) constituted an insignificant part of the sample. Finally, we assumed that movement patterns, survival rates and productivity of Western Gulls from other (non-SEFI) northern California colonies, which constituted 15% of the northern California population (Sowls et al. 1980), were not significantly different from that of Farallon gulls (see Coulter 1973, Briggs 1977, Bellrose 1983).

Survival Rates of Breeding Adults

Between 1978 and 1986, we monitored 231 individually marked adults during 611 gull-years as they returned to their breeding territories from year to year. Sexes of breeding adults were determined by observing copulations, courtship feeding of females, or size differences; males are larger than females, there being little overlap in weight between the sexes (Pierotti 1981). We have found no evidence for female-female pairing at SEFI (see Results; Sex-ratios among Independent Birds). Except for 28 birds monitored in 1978–1979 and 1979–1980, ages of each were known and included three- to 19-year-old gulls. All known-age adults had been banded as young with an individually numbered USFWS aluminum, or Lambournes incoloy, band on one leg and a polyvinyl chloride (Warner) color band on the other. A different color or leg combination has been used each year. Because of the high loss rate for aluminum bands after the third year (Spear 1980), birds that had received them were rebanded with highly durable stainless steel or incoloy bands prior to monitoring for survival rates. Each gull also received a new Warner band for which the loss rate is low (Spear 1980). Bands were color-coded with strips of plastic 3M Scotch tape to further facilitate recognition of individuals. In 1986, this tape was still present on nearly all bands.

Farallon gulls rarely skipped a year of breeding after recruiting into the breeeding population; birds that returned to the colony failed to breed in only eight of the 505 gull-years monitored (for reasons given below, 106 birds that failed to return are excluded). Nevertheless, with the exception of one bird, these nonbreeding gulls continued to maintain their territories or to frequent former territory locations. Each year we made band reading surveys throughout the colony in search of birds that no longer occupied their known territories to determine if they had established territories elsewhere. In only four instances did gulls select new nest sites that were more than 10 m from the original site. From 1978 through 1984, LBS also read bands of 2287 different adults at SEFI clubs and along the California, Oregon and Washington coasts. None of the adults that no longer occupied breeding territories was among them. During the course of the study, only one bird that was not seen in the year following a breeding year was seen subsequently in the colony. In view of the above, and

TABLE 1

ESTIMATED SURVIVAL RATES OF THE SEXES OF FIRST-, SECOND-, AND THIRD-YEAR WESTERN GULLS BASED ON COASTAL CENSUSES OF DEAD BIRDS, 1978–1986[a]

Age class	No. of birds found dead on coast (N_i)		No. of birds at risk (R_i)		Survival rates	
	Male	Female	Male	Female	Male	Female
First-year	213	179	426	459	0.50	0.61
Second-year	54	44	213	280	0.75	0.84
Third-year	29	25	159	236	0.82	0.89

[a] For method of calculating survival rates, see Methods: Survival Rates of First-, Second-, and Third-year Birds. For these calculations, N_a, the number of adults of breeding status found dead on the coast was 108 males, i.e., 16% less than the total number found (see Table 8) due to adjustment to exclude nonbreeding adults (see Results: Population Status). The mean annual mortality rate of breeding males (M_a) was 16% (see Table 3). The mean productivity rate (P) for the period 1978–1986 was 0.66 young per breeding adult (PRBO, unpubl.). Because of evidence indicating a female excess among fledglings (see Results: Sex Ratios of Young at Dispersal), the productivity figures used here were 0.630 for males, and 0.685 for females, i.e., 0.92 M/F produced.

the lack of emigration to other colonies, we assumed that all adults disappearing from the colony had died.

Life expectancy after the onset of breeding was estimated from the relationship

$$\frac{\sum_{i}^{n} [(N)(i)]}{\sum_{i}^{n} (N)},$$

where (i) is the number of years that a gull has bred, and (N) is the number of gulls surviving in the respective year of breeding (i). Our formula differs from that used by Lack (1954) because we consider age-related variation in survival rates, whereas Lack assumed a constant rate (see also Coulson and Wooller 1976).

Number and Sex Ratio and Nonbreeding Adults

Breeding adults rarely attended clubs during the breeding season. During 1978 through 1984, LBS identified 980 breeding adults by reading their bands while they occupied breeding territories on SEFI. During the same period, he read bands of 623 adults at SEFI clubs; these included only 9 (1.4%) of the 980 identified breeders. Therefore, we considered the number of breeding adults at island clubs to be insignificant compared to the number of nonbreeders.

During the breeding season in all years from 1978 through 1986, counts of Western Gulls attending SEFI clubs were made during evening when numbers were highest. We estimated age-ratios of nonbreeding adults by censusing known-age (color-banded) birds, which composed 8–12% of the Farallon gull population), and adjusting counts of respective age classes for band loss (see Spear 1980). To determine sex ratios of nonbreeding adults, LBS first identified individual gulls at clubs by using a spotting scope to read the numbers on their metal bands. He then sexed 79% of these birds by comparing size differences; this sexing method has an accuracy of 95%, and error is not biased toward either sex (Spear 1981; see also Pierotti 1981, Hand 1986).

Mortality Factors in Chicks

During 1979 through 1986, we collected 658 dead young on SEFI and examined each, noting sex, body condition, and injuries or other ailments. Since the leg muscles of healthy, prefledged young are well developed by 10 d of age (but pectoral muscles are not), we considered a prefledged chick to have starved only if it had atrophied leg muscles, a criterion based on examination of chicks known to have starved. We assumed that young found emaciated as well as seriously pecked (the result of entering territories of non-related adults) had starved before they were pecked (see Fordham 1970, Ward 1973, Hunt and Hunt 1975).

Mortality Factors on the California Coast

LBS examined 919 Western Gulls found dead along the northern coast of California, noting age, sex, body condition, and injuries or other ailments. Each was skinned and examined subcutaneously for wounds on the head and body. Examination of internal organs included opening and inspection of stomachs and proventriculi. Birds with no visible injuries or disease, but having atrophied pectoral muscles, were considered to have starved. Gulls with ailments that were not readily identified were frozen and later examined by veterinarians. We considered mortality factors of Western Gulls within the study area as representative of Farallon gulls because: (1) 85% of the Western Gulls that bred within this area bred on SEFI (Sowls et al. 1980), and most remain within the area (Coulter 1975); and (2) the number of Western Gulls from outside populations that move into this area is low (see above).

Mortality Factors in Adults at SEFI

Between 1978 and 1985, we collected 496 dead adults on SEFI during all seasons and examined each as described above for gulls found dead on the California coast. In 1983 and 1984, 33 adults that had died on SEFI during the breeding season (April through August) were collected at random and sent to the USFWS National Wildlife Laboratory (Madison, Wis.) for examination.

RESULTS

Sex Ratios of Young at Hatching and Dispersal

The sex ratio of 99 Farallon Western Gulls examined at hatching was 0.98 M/F. We did not

TABLE 2
SEX RATIOS AMONG PREBREEDING WESTERN GULLS FOUND DEAD ON THE NORTHERN CALIFORNIA COAST DURING FALL AND WINTER CENSUSES, 1978–1986

	First-year gulls		Subadults	
	Mean no. found per census	Sex ratio M/F	Mean no. found per census	Sex ratio M/F
1978–1979	26.0	1.17	13.0	1.60
1979–1980	48.5	1.26	12.0	1.18
1980–1981	26.0	1.26	12.5	0.92
1981–1982	27.5	1.20	9.0	1.25
1982–1983	61.0	1.35	14.5	1.07
1983–1984	20.5	0.95	11.0	1.44
1984–1985	23.0	1.09	7.0	1.80
1985–1986	19.5	0.95	8.0	1.00

TABLE 3
THE ANNUAL SURVIVAL RATES OF MALE AND FEMALE WESTERN GULLS BREEDING AT SOUTHEAST FARALLON ISLAND, 1978–1986[a]

At risk from-to	Males	Number monitored	Females	Number monitored
1978–1979	0.84	25	0.83	23
1979–1980	0.85	20	0.74	19
1980–1981	0.89	26	0.86	21
1981–1982	0.82	22	0.83	18
1982–1983	0.80	20	0.72	18
1983–1984	0.83	59	0.77	53
1984–1985	0.85	74	0.89	64
1985–1986	0.86	80	0.85	69
All years	0.84	326	0.81	285

[a] Survival rate estimates based on observations of banded breeding gulls monitored at SEFI (see Methods: Survival Rates of Breeding Adults).

find significant differences in the sex ratio among young hatched from first- (M/F = 0.74), second- (M/F = 1.20) or third-laid eggs (M/F = 1.06) (G test, $P > 0.05$), however, these results should be considered as preliminary because of small sample sizes. The SEFI ratios are similar to those of Western Gulls hatching in a southern California colony (Sayce and Hunt 1987), but opposite to results of Ryder (1983) who found a strong correlation between egg sequence and the sex of Ring-billed Gull (*L. delawarensis*) chicks, in which more males hatched from first eggs while more females hatched from second eggs.

The sex ratio of 658 chicks found dead during all phases of the chick period was 1.08 M/F. Sayce and Hunt (1987) also found evidence for a small excess of male chick mortality (0.89 M/F alive at fledging, n = 1291), apparently due to factors similar to those found during this study (see Results: Mortality Factors of Chicks). Although the M/F ratio in our sample was not statistically different from 1.0 (G test, $P > 0.05$), we assumed that the sex ratio of young dispersing from the Farallons colony was 0.92 M/F.

SURVIVAL RATES AMONG INDEPENDENT BIRDS

Estimates of mean annual survival rates of first-, second-, and third-year birds are given in Table 1. Females had significantly higher survival rates than males (G test; first-year, $P < 0.001$; second- and third-years, $P < 0.01$). The ratio of first-year males to females in coastal samples was highest in years when the greatest numbers of first-year birds were found ($r = 0.777$, $df = 6$, $P < 0.05$; Table 2). This indicates that when mortality was high it increased faster among males than females. This relationship was not found among second- and third-year gulls (subadults) $r = -0.338$, $df = 6$).

Annual survival rates of breeding adults are given in Table 3. Males had higher survival rates than females in six of eight years, although the difference was not significant (t test, $P < 0.2$). The regression of annual male survival rate (X) on female survival rate (Y) gives the relationship $Y = 1.354X - 0.330$. If male and female survival rates changed similarly, the slope of the regression should be 1.0, whereas it is 1.35, indicating a tendency (t test $P < 0.1$) for female survival rates to decrease more rapidly in years when survival is low. Thus, in years of greater mortality rates, the effect is proportionally greater among females. The regression indicates that survival rates of the sexes would be equal when the overall survival rate is 0.93.

Both sexes of first-year birds had significantly lower survival rates than respective sexes of second-year birds (G test, $P < 0.001$, Table 1). Survival rates of second-year males was significantly lower than that of adult males, and third-year females had significantly higher survival rates than adult females (G test, $P < 0.01$; see Table 3 for SR values of adults). Survival rates of younger breeding adults (3–8 years old) averaged higher than that of older breeders (9–19 years old), but the difference was not significant (t test; $P < 0.2$, Table 4). The tendency among different age classes of breeding birds is similar to that found in Kittiwakes (*Rissa tridactyla*), in which younger birds had significantly higher survival rates than older birds (Coulson and Wooller 1976).

SEX-RATIOS AMONG INDEPENDENT BIRDS

The age/sex-related survival rates indicate a female bias in sex ratios of each age class from one to 14-year-old birds (Table 5). The greatest female bias in sex ratios is expected for three- and four-year-old cohorts. Assuming that survival rates of gulls older than 19 years are similar to those of 9–19-year-olds, we estimated the sex

TABLE 4
THE ANNUAL SURVIVAL RATES OF "YOUNG" AND "OLD" WESTERN GULLS BREEDING ON SOUTHEAST FARALLON ISLAND, 1978–1986[a]

Years of age	Males	Gull yrs. monitored	Females	Gull yrs. monitored
Young (3–8 years)[b]	0.85	118	0.84	129
Old (9–19 years)	0.83	138	0.79	126

[a] Survival rate estimates based on observations of banded breeding gulls monitored at SEFI (see Methods: Survival Rates of Breeding Adults).
[b] These data include 6 gull-years of males that bred in their third year.

TABLE 5
SEX RATIOS AMONG DIFFERENT AGE CLASSES OF FARALLON WESTERN GULLS ESTIMATED FROM SURVIVAL RATES OF RESPECTIVE GROUPS (1978–1986)[a]

Age	Percent surviving: Males	Percent surviving: Females	Sex ratio M/F
1	50.0	61.0	0.82
2	37.5	51.2	0.73
3	30.8	45.6	0.68
4	26.2	38.3	0.68
5	22.3	32.2	0.69
6	19.0	27.0	0.70
8	13.8	19.1	0.72
10	9.5	11.9	0.80
12	6.6	7.4	0.89
14	4.6	4.6	1.00
16	3.2	2.8	1.14
18	2.2	1.7	1.29
20	1.5	1.0	1.50
22	1.0	0.6	1.67
24	0.7	0.4	1.75

[a] Survival rates are calculated from coastal censuses of dead birds for age classes 1–3, and from observations of banded breeding adults for all remaining age classes (see Methods: Survival Rates of First-, Second-, and Third-year Birds, and Survival Rates of Breeding Adults).

ratio among the adult population at 0.77 M/F. This is similar to the female bias (0.67 M/F) observed among adults in another population of Western Gulls (Hunt et al. 1980).

In spite of a large excess of females in the adult population, we found no evidence of a bias in the sex ratio of breeding birds and no evidence of female-female pairing. Only one of the many thousands of nests examined over nine years contained a supernormal clutch (4 eggs), and we have not seen pairs of two female-sized birds. We assumed, therefore, that the sex ratio among breeding birds was unity, and expected to find a female excess among nonbreeders. Because the age that male Farallon gulls breed for the first time averages younger than that of females (M = 4.7 years, F = 5.6 years; Spear, unpubl.), we expected a sex ratio in the population of nonbreeding adults of 0.32 M/F. Within SEFI clubs, however, we found a female bias of lesser magnitude (0.48 M/F). This included 0.63 M/F for four-year-olds, and 0.35 M/F for birds >4 years. The discrepancy between predicted and observed values may have resulted from a greater tendency for nonbreeding females to remain away from SEFI compared to nonbreeding males (Spear 1981).

POPULATION STATUS

The evidence suggests that the number of gulls that attain breeding age should nearly equal that of adults that die, i.e., that the population status is relatively stable (Table 6). This conclusion is corroborated by the low incidence of emigration and stability in the breeding population. Limitations imposed by a shortage of unused breeding space could also be responsible for stability, however, in which case the status of the nonbreeding adult population should be the best indication of the population status as a whole. The censuses of nonbreeding adults in clubs show a gradual increase from 1978 through 1983, an abrupt decline in 1985, and partial recovery in 1986 (Fig. 1). The decline in 1985 apparently resulted from heavy recruitment of nonbreeders into the breeding population following two years of higher-than-average mortality among breeders (Table 3). These data indicate that the nonbreeding population has been only slightly greater than that required to replace breeding adults that have died (approx. 4550 per annum), and that on a long-term basis the nonbreeding population is relatively stable.

Based on the above, and coastal censuses showing small numbers of adults more than one foraging range (85 km; Hunt et al. 1979) from SEFI during the breeding season (LBS and DGA, ms), we estimate that the nonbreeding adult population averaged approximately 7500 birds per annum (1978–1986); i.e., composed 22% of the

TABLE 6
EXPECTED RATES OF REPLACEMENT OF BREEDING ADULT WESTERN GULLS AT SOUTHEAST FARALLON ISLAND

	Survival rate: fledging to recruitment[a]	Subsequent no. breeding years expected	Young dispersed/ breeding adult/year[b]	No. progeny surviving to breeding age/adult dying
Males	0.24	6.2	0.630	0.94
Females	0.29	5.2	0.685	1.03

[a] The mean age of recruitment of males is approximately 4.7 years, and of females, 5.6 years (Spear, unpubl.).
[b] This value is the mean for the period of 1978–1986 (PRBO; unpubl., this study).

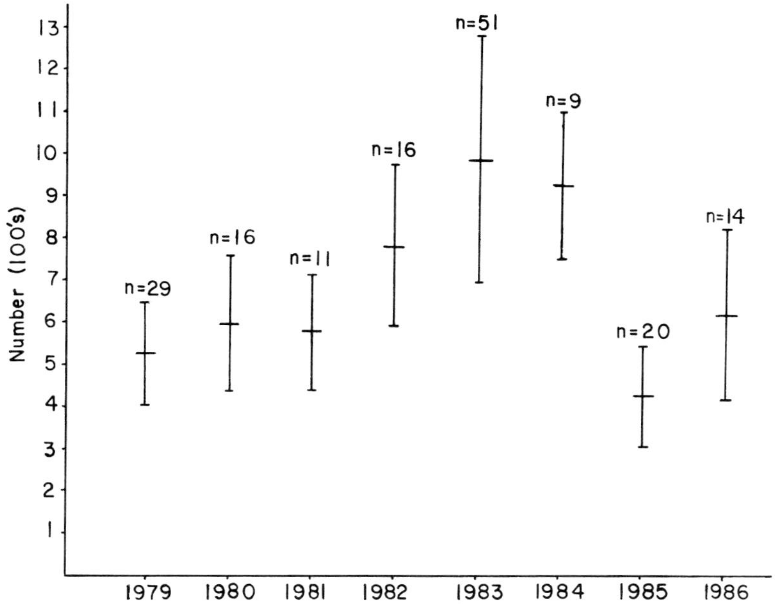

FIGURE 1: Evening counts of nonbreeding adult Western Gulls at clubs on Southeast Farallon Island, 1979–1986; "n" denotes the number of counts, central bar shows the mean, and bars above and below the mean show one standard deviation on each side of the mean.

adult population. Based on a sex ratio of 0.48 M/F in the population of nonbreeding adults (see preceding section), the adult male population (including an estimated 13,000 breeders) would include approximately 2440 (16%) nonbreeders (i.e., just enough males to replace breeders that die), and the adult female population, 5060 (28%) nonbreeders.

Mortality Factors among Chicks

The relative importance of mortality factors among chicks varied significantly during different months (G test, $P < 0.001$, Table 7; gulls dying from unknown causes were excluded from this analysis). In June, when most young hatched (Coulter 1973), starvation caused at least 50% of the deaths. In July, when young were highly mobile and beginning to fly, 54% of the deaths were due to pecking by adults. During the dispersal period of August and September, 91% of the dead young had starved. More male chicks starved when compared to females, while more females died from being pecked by adults (G tests, $P < 0.05$).

Conspecific predation is not usually common on SEFI, but appeared to be an important mortality factor during the El Niño year of 1983 when natural food was scarce and many chicks from one to 10 d old were eaten. We could not examine these chicks and assumed that predators were not sexually selective.

Mortality Factors on the California Coast

Starvation was a more important mortality factor among first-year gulls than among subadults (G test, $P < 0.001$, Table 8); 59% of the mortality of first-year birds was caused by this factor. The relative importance of different mortality factors did not vary significantly between subadults and adults (G test, $P > 0.05$, sexes were grouped and variables tested were starvation, disease, and combined accidents); most were accidental, with gunshot wounds ranking first. Broken wings, many of which probably resulted from gunshot, were also a major factor.

Starvation was a more important mortality factor among first-year males than among first-year females (G test, $P < 0.001$; Table 8). In contrast, there was a higher incidence of starvation among subadult and adult females (grouped) than respective age classes of males (G test, $P < 0.05$). During a given year, the percent of first-year males dying from starvation increased significantly as the number of first-year gulls found dead increased ($r = 0.727$, $df = 6$, $P < 0.05$; Fig. 2), and a similar relationship existed among subadult and adult females (grouped) $r = 0.792$, $df = 6$, $P < 0.02$; Fig. 3). A similar tendency among subadult and adult males was not significant ($r = 0.544$, $df = 6$, $P < 0.2$).

The number of subadults and adults (grouped) that died from accidental causes was highest in

TABLE 7
MORTALITY FACTORS IN 658 WESTERN GULL CHICKS ON SOUTHEAST FARALLON ISLAND, 1978–1986

	N	Starvation: Starved only (%)	Starvation: Starved & pecked (%)	Starvation: Total (%)	Pecked only (%)	Other (%)[a]	Unknown (%)
June							
Males	42	36	17	53	28	2	17
Females	36	35	11	46	31	6	17
July							
Males	133	14	22	36	53	0	11
Females	135	10	22	32	56	2	10
August							
Males	124	19	46	65	12	10	13
Females	120	21	36	57	23	5	15
September							
Males	43	93	0	93	0	7	0
Females	25	88	0	88	0	12	0
Overall							
Males	342	29	27	56	29	4	11
Females	316	23	25	48	36	4	12

[a] "Other" causes of mortality were broken wings, deformities, and entrapment within or between rocks, vegetation, debris and buildings.

years when the number dying from starvation was highest (males; r = 0.816, df = 6, $P < 0.02$, and females; r = 0.793, df = 6, $P < 0.02$, Fig. 4). Other evidence suggesting that starvation and accident rates were interrelated was the significant correlation of annual survival rates of adults (mortality due mostly to accidents) with that of first-year gulls (mortality due mostly to starvation) (r = 0.924, df = 6, $P < 0.002$, Table 9).

MORTALITY FACTORS IN ADULTS AT SEFI

The relative importance of mortality factors in adults found dead at SEFI varied significantly between seasons (G test, $P < 0.001$; sexes were grouped and variables tested were starvation, disease and combined accidents, Table 10). Lower rates of starvation occurred during spring, and lower accident rates during summer. The relative importance of mortality factors did not vary significantly between the sexes (G test, $P > 0.05$; seasons grouped with variables same as above). We could determine the cause of death, however, in only 36% of the cases. The cause of many undiagnosed deaths was probably Botulism C toxin. During all years we noticed that a large proportion of the gulls dying on SEFI during the

TABLE 8
MORTALITY FACTORS IN 919 WESTERN GULLS FOUND DEAD DURING ALL SEASONS ON THE COAST OF NORTHERN CALIFORNIA, 1978–1986, AND THEIR PERCENT VALUES

	N	Starved	Diseased[a]	Accidents: Broken wing	Accidents: Fishing gear	Accidents: Shot	Accidents: Other[b]	Accidents combined	Unknown
First-year									
Males	260	68	0	13	3	4	8	28	4
Females	218	49	0	19	3	10	11	43	8
Subadults									
Males	88	23	3	17	8	29	17	71	3
Females	73	30	0	20	6	24	14	64	6
Adults									
Males	128	14	7	15	8	30	22	75	4
Females	153	27	6	24	1	24	14	63	4

[a] Diseases included gout, tuberculosis, and sinusitis. The incidence of aspergilosis was 70%, 68%, and 76%, respectively, among emaciated adults, subadults, and first-year gulls. Because this disease is a secondary infectant we did not consider it a mortality factor.

[b] Most common "other" accidents included plastic six-pack wrappers over head, poisoning, collision with automobiles and wires, entrapment in fish bins, oiling, and entanglement in machinery.

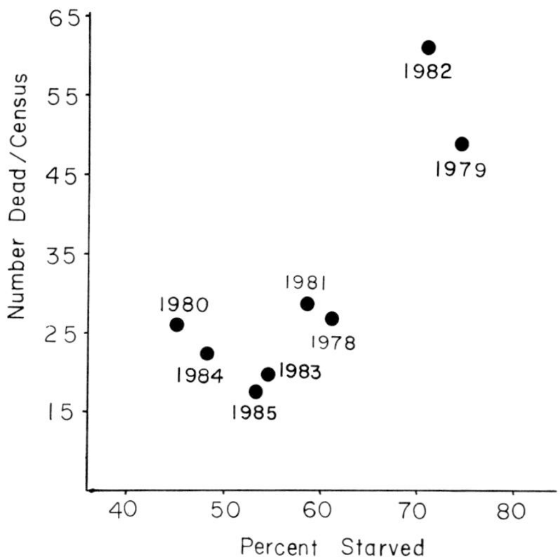

FIGURE 2. Percent of first-year males dying from starvation as a function of the number found dead during coastal surveys, 1978–1986.

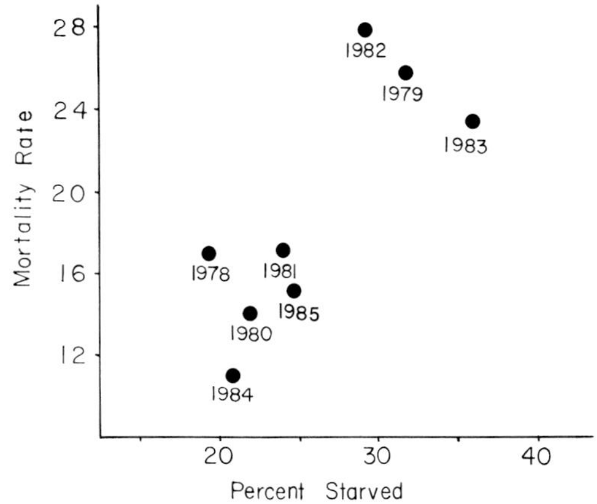

FIGURE 3. Percent of subadult and adult females (grouped) that died on the mainland coast from starvation as a function of the mortality rates of breeding females monitored on Southeast Farallon Island, 1978–1986.

breeding season developed symptoms of botulism. Tests on 33 gulls by the USFWS confirmed this; at least 42% of the gulls examined died from Botulism C toxin.

Mortality during the breeding season was not severe, however. Among 65 breeding adults that died while being studied during the five years including 1978–1980 and 1982–1985 (see Table 3), only 13%, 0%, 33%, 9%, and 0% ($\bar{x} = 11 \pm 12.1\%$), respectively, died during the spring and summer. Of the deaths that occured per annum, the proportion occurring during the breeding season varied significantly between years (G test, $P < 0.05$), being highest in years (1978, 1983) when food was apparently least abundant, i.e., in years when productivity of gulls and other seabirds breeding on SEFI was lowest (PRBO, unpubl.). Years of exceptionally low food availability are infrequent, however, as only three such years have occurred during the last 16.

Chronology of Coastal Mortality

Recovery rates of dead Western Gulls on the northern California coast peaked during winter in all age classes, although that of first-year gulls and adults was bimodal, with smaller peaks occurring in the fall (Fig. 5).

Additional evidence suggests that these patterns are realistic: This includes: (1) close agreement between the percent of dead adults we found on the coast during the breeding season and the percent of annual deaths in that same period among known adults monitored for survival on SEFI; 12% on the coast vs. 11% on SEFI; and (2) a significant correlation in numbers of dead adults found per annum during coastal searches with that of mortality rates among known adults monitored on SEFI ($r = 0.735$, $df = 6$, $P < 0.05$, Table 9).

The mortality patterns in Figure 3 are different from ones derived from band recoveries, which show peak recoveries in late summer (Coulter 1975). These differences may have resulted from several factors. First, the general public is most active during spring and summer, especially at

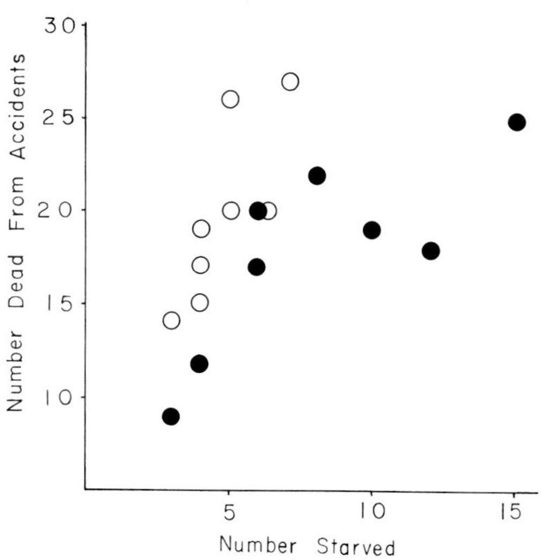

FIGURE 4. Mortality on the mainland coast: number of subadult and adult Western Gulls (grouped) that died from accidents as a function of the number dying from starvation, 1978–1986. Solid circle = females, open circle = males.

TABLE 9

THE ESTIMATED ANNUAL SURVIVAL RATES OF FIRST-YEAR WESTERN GULLS FROM SOUTHEAST FARALLON ISLAND, 1978–1986[a]

Year	No. of 1st-yr. gulls found dead (N_1)	No. of breeding adults found dead (N_a)	Young dispersed/ breeding adult/year[b] (P)	Survival rate of adults[c] ($1 - M_a$)	Est. survival rate of 1st-yr. gulls (SR^{0-1})
1978–1979	22	16	0.52	0.83	0.55
1979–1980	79	31	0.81	0.79	0.34
1980–1981	42	28	0.72	0.87	0.73
1981–1982	45	27	0.54	0.83	0.48
1982–1983	100	39	0.90	0.76	0.32
1983–1984	34	33	0.33	0.80	0.38
1984–1985	38	20	0.67	0.87	0.63
1985–1986	32	23	0.76	0.85	0.73

[a] See Methods: Survival Rates of First-, Second-, and Third-year Birds, for method of calculating survival rates.
[b] PRBO (unpubl. data).
[c] Survival rate values ($SR = 1 - M_a$) represent the mean between males and females (Table 8).

beaches where, during those respective seasons, we found the majority of dead gulls (see also Kadlec and Drury 1968). Moreover, during fall and winter, when the public is less active, we found the majority of dead gulls at dumps. Even if search efforts by the public were consistent over time, people are less likely to examine a dead gull in a dump than a dead gull on the beach. Another bias for adult gulls may have resulted from using band recoveries from SEFI, especially those from August and September when biologists regain access to the entire island. (Activities are restricted near to paths during the breeding season.) Many birds that died throughout the breeding season are then inspected, and this may have biased band recovery rates towards late summer. For example, we found nearly twice as many dead adults on SEFI during the breeding season than we found on the coast during all seasons, although only 11% of the mortality of known breeding adults occurred during the breeding season. Third, the temporal pattern of mortality may have changed since the period (1933–1971) when the band recovery information used by Coulter (1975) was accumulated.

DISCUSSION

When production of one sex is more costly than another, selection should favor the production of unequal sex ratios (reviewed by Meyers 1978). In applying this hypothesis to species of birds which are sexually dimorphic in size (such as Western Gulls; Pierotti 1981) and that breed in food-limited environments, Meyers predicted that the larger sex should have lower survival rates due to starvation in the nestling stage. This

TABLE 10

MORTALITY FACTORS AMONG 496 ADULTS FOUND DEAD AT SOUTHEAST FARALLON ISLAND, 1978–1985, AND THEIR PERCENT VALUES

	Gulls examined (N)	Starved	Diseased[a]	Accidents: Broken wing	Accidents: Shot	Accidents: Other[b]	Accidents combined	Unknown
Fall–winter								
Male	55	24	9	0	0	16	16	51
Female	57	37	5	5	2	9	16	42
Spring								
Male	40	17	5	3	2	10	15	63
Female	68	7	3	1.5	3	1.5	6	84
Summer								
Male	135	25	4	0	0	1	1	70
Female	141	30	5	0	1	1	2	63
Total								
Male	230	23	5	<1	<1	6	<8	64
Female	266	26	4	1	2	3	6	64

[a] Diseases included gout, tuberculosis, and sinusitis.
[b] "Other accidents" included oiling, broken beak, and entanglement with fishing tackle and six-pack wrappers.

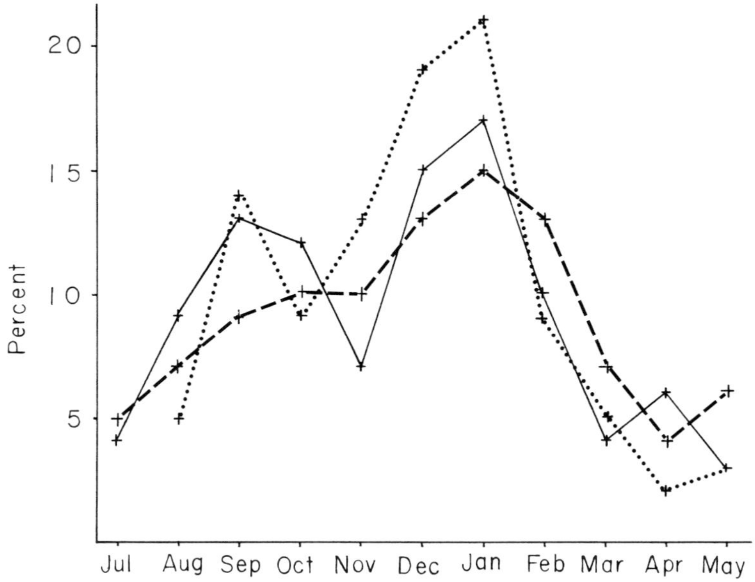

FIGURE 5. Chronology (by month) of mortality in three age classes of Western Gulls on the northern California coast, 1978–1986. Solid line = adults (4 yrs and older), dashed line = subadults (second- and third-year birds), and dotted line = first-year birds. Values are given as the mean number of birds found per monthly census expressed as the percent of the mean annual total (no coastal searches in June).

was demonstrated in icterids and corvids (reviewed by Meyers 1978, Roskaft and Slagsvold 1985) but was not found for other species (reviewed by Sayce and Hunt, 1987). Nevertheless, before assessing the importance of selective factors operating differentially upon the sexes, survival rates and mortality factors among all age groups must be studied (Meyers 1978). Our study is the first to attempt this in an avian species. We have found additional support for Meyers' prediction, as well as evidence that other selective factors act differentially on sexes of all age groups. In the following discussion these various factors are evaluated.

The evidence indicates that food available to Farallon gull chicks was often in limited supply. During this study at least 29% of all young hatched, died before dispersing (PRBO, unpubl.). Since 52% of the dead chicks had starved (Table 7), it follows that at least 15% of all young hatched died from this factor. Similarly, data from Coulter (1977) indicated that 19% of the Western Gulls hatched on SEFI during 1969–1970 died as a result of "food limitation." The higher rate of starvation among male chicks does support Meyers' (1978) prediction, but interestingly, did not result in a significantly biased sex ratio prior to dispersal because of higher mortality rates among females from attacks by adults. Death from pecking (by unrelated adults) was most frequent in July when recently fledged young (with poor flight control) frequently landed on territories other than their own. Fledgling females average smaller in size than males (Sayce and Hunt 1987), and thus their greater mortality during fledging may have resulted from being less able to withstand adult attacks as they returned to their territories. In colonies of low density this may not be an important mortality factor because, with larger territories, young have a better chance of landing on their own territory. This could explain a smaller female bias in the sex ratio of Farallon young at dispersal (0.92 M/F) compared to that of Western Gull young on Santa Barbara Island, California (0.89 M/F; Sayce and Hunt 1987), where territories average 15 times larger than territories on SEFI (Hunt and Hunt 1975, Pierotti 1981). Nevertheless, the higher starvation rate among male chicks on SEFI suggests that many may have been in poorer physical condition than females when they dispersed, and this could, in part, explain the higher mortality rates of first-year males after dispersal. Possible advantages gained by females during the chick period, however, probably diminish by fall. When mortality peaked during the winter, survival was probably most dependent on foraging skills. It is likely, therefore, that lower survival rates of first-year males resulted primarily from their need to meet higher energy demands (due to larger body size) before foraging skills were sufficiently developed (see also Sayce and Hunt

1987). (For review of age-related foraging success, see Briggs 1977, Verbeek 1977, Searcy 1978, Burger 1981.) In some sexually diomorphic species the larger sex also fledges at a lower percent of the adult weight (Ricklefs 1968, Bancroft 1984). As a result, independent young of the larger sex are at a disadvantage; besides meeting higher energy demands for maintenance they must grow more, proportionally, to attain adult size.

Although nearly 50% of the deaths of independent males less than four years old occurred during the first year as the result of starvation, the greatest biasing of the sex ratio occurred during the second and third years due to accidents involving humans. It is important to note, therefore, that accident and starvation rates were correlated. This suggests that food limitation was a more important factor than indicated by starvation rates alone; gulls are probably bolder when hungry. But why would subadult males be more prone to death from starvation and accidents than females if foraging skills were improved and, because of larger size, they held an advantage over females when competing for stationary food sources (i.e., human refuse; see Briggs 1977, Monaghan 1980)? One possibility is that males may be bolder than females and thus more prone to accidents. Another possibility is that this could be related to the tendency for subadult females to remain farther away from SEFI than subadult males (Spear 1981). The majority of subadult males remain within foraging range of adults and may, therefore, be more prone to higher stress related to competition for food.

The tendency for a higher survival rate of breeding males compared to breeding females (Table 3) is interesting. Coulson and Wooller (1976) found higher survival rates among female Kittiwakes and suggested that greater mortality among males resulted from stress associated with territory establishment. Although territory establishment is also primarily the responsibility of male Western Gulls (Pierotti 1981, Hand 1986), the tendency for lower survival rates of Farallon females was probably due to differences in ability to make use of local foraging opportunities. Pierotti (1981) found that females caring for young take longer than males to complete foraging trips and return with lighter food loads. The reason is not known, but this suggests that females would be at a disadvantage during times of food shortage. Competition for food among adults did, in fact, appear severe during late fall and winter of some years when many foraged up to 70 km from SEFI at San Francisco Bay Area (SFBA) dumps (Spear 1979). At this time, most adults forage in the vicinity of SEFI to maintain territories (males) and mates (females). During these period, females from SEFI had to compete for a stationary food source not only with male Western Gulls, but also with even larger Glaucous-winged Gulls (*L. glaucescens*) which compose up to 50% of the large larids foraging in the SFBA during winter (Cogswell 1969, Spear unpubl.). Moreover, competition in the SFBA has probably increased between 1979 and 1986 due to the closure of three of the six dumps most important to large larids. The high winter mortality of adults and sharper decrease in survival rates with increases in the incidence of starvation of SEFI females suggests that the above discussion is applicable. It is interesting that in British populations of Herring Gulls that were increasing at 12% per annum (i.e., little evidence of food shortage), no difference was found between survival rates of the sexes (Chabrzyk and Coulson 1976, but see Coulson and Butterfield 1986). These gulls were experiencing a survival rate of 0.93, equal to the projected value for Farallon gulls when survival rates of the sexes should be equal.

In conclusion, it appears that sexual differences in the survival rates of Farallon gulls are determined primarily by three factors: higher rates of starvation among first-year males, higher incidence of accidents among subadult males (amplified by factors related to food shortage), and a tendency for higher rates of starvation in breeding females. If larid populations are, in fact, often limited in size by food (see Lack 1954, Ashmole 1971, Furness and Birkhead 1984, Coulson and Thomas 1985), our results suggest that a shortage of males can be expected in many populations and that this may not be a recent phenomenon. This could explain instances of female-female pairing and polygamy in these otherwise monogamous species (Conover and Hunt 1984, Coulson and Thomas 1985), although other factors may affect sex-ratios in some populations (see Fry and Toone 1981; Fry et al., 1987). In the case of the SEFI population, the absence of female-female pairs may result from high density nesting where the smaller females cannot compete with males for breeding territories (Hand 1980, Conover and Hunt 1984).

ACKNOWLEDGMENTS

We thank the San Francisco chapter of the Oceanic Society for transporation to and from SEFI. The USFWS allowed our research on the Farallon National Wildlife Refuge and provide invaluable assistance in examining mortality factors; we especially thank R. Lowe and R. K. Stroud, D.V.M. Other persons who advised on mortality factors or examined gulls include D. M. Fry, L. Lowenstein, D.V.M. (University of California, Davis), and P. Quinn (Monterey County SPCA). Dr. J. Martin (Moss Landing Marine Laboratories) and M. R. Graybill, (Oregon Institute of Marine Biology) provided LBS with facilities for examining dead gulls he found on the coast. We are grateful to them, and to R. J. Boekelheide, R. P. Henderson and the many Farallon volunteers who collected carcasses on SEFI. Logistical support for coastal work was provided by J. C. and E. B. Spear, and S. Goldhaber kindly typed the final draft. This paper was much improved by the comments of M. C. Coulter, D. M. Fry, J. L. Hand, G. L. Hunt, Jr., and C. A. Ribic. Spear was funded in part by a Packard Foundation grant administered through Moss Landing Marine Laboratories. This is PRBO contribution no. 330.

LITERATURE CITED

AINLEY, D. G., AND T. J. LEWIS. 1974. The history of Farallon Island marine bird populations, 1854–1972. Condor 76:432–446.

ANDERSON, D. R., K. P. BURNHAM, AND G. C. WHITE. 1985. Problems in estimating age specific survival rates from recovery data of birds ringed as young. J. Anim. Ecol. 54:89–98.

ASHMOLE, N. P. 1971. Seabird ecology and the marine environment. Pp. 223–286 *in* D. S. Farner and J. R. King (eds.). Avian biology, vol. 1. Academic Press, New York.

BANCROFT, G. T. 1984. Growth and sexual dimorphism of the boat-tailed Grackle. Condor 86:423–432.

BELLROSE, C. A. 1983. The breeding biology and ecology of a small mainland colony of Western Gulls (*Larus occidentalis*). Unpubl. M.S. thesis, San Jose St. Univ., Calif.

BOTKIN, D. B., AND R. S. MILLER. 1974. Mortality rates and survival of birds. Am. Nat. 108:181–192.

BRIGGS, K. T. 1977. Social dominance in young Western Gulls: its importance in survival and dispersal. Unpubl. Ph.D. dissertation, Univ. Calif., Santa Cruz.

BURGER, J. 1981. Feeding competition between Laughing Gulls and Herring Gulls at a sanitary landfill. Condor 83:328–335.

CHABRZYK, B., AND J. C. COULSON. 1976. Survival and recruitment in the Herring Gull, *Larus argentatus*. J. Anim. Ecol. 45:187–203.

COGSWELL, H. L. 1969. Gulls and solid waste disposal in the San Francisco Bay Area, California. Proc. World Conf. Bird Hazards to Aircraft, pp. 421–439. Kingston, Ont., Sept. 1969.

CONOVER, M. R., AND G. L. HUNT, JR. 1984. Experimental evidence that female-female pairs in gulls result from a shortage of breeding males. Condor 86: 472–476.

COULSON, J. C., AND E. WHITE. 1959. The post-fledging mortality of the kittiwake. Bird Study 6:97–102.

COULSON, J. C., AND R. D. WOOLLER. 1976. Differential survival rates among breeding Kittiwake gulls, *Rissa tridactyla*. J. Anim. Ecol. 45:205–213.

COULSON, J. C., AND C. S. THOMAS. 1985. Changes in the biology of the kittiwake (*Rissa tridactyla*): a 31-year study of a breeding colony. J. Anim. Ecol. 54:9–26.

COULSON, J. C., AND J. BUTTERFIELD. 1986. Studies on a colony of colour-ringed Herring Gulls *Larus argentatus*: adult survival rates. Bird Study 33:51–54.

COULTER, M. C. 1973. Breeding biology of the Western Gull, *Larus occidentalis*. Unpubl. M.S. thesis, Oxford Univ.

COULTER, M. C. 1975. Post-breeding movements and mortality in the Western Gull, *Larus occidentalis*. Condor 77:243–249.

COULTER, M. C. 1977. Growth, mortality, and the third chick disadvantage in the Western Gull, *Larus occidentalis*. Unpubl. Ph.D. dissertation, Univ. of Pennsylvania, Philadelphia, Penn.

FORDHAM, R. A. 1970. Mortality and population change of Dominican Gulls in Wellington, New Zealand, with statistical appendix by R. M. Cormak. J. Anim. Ecol. 39:13–27.

FRY, D. M., AND C. K. TOONE. 1981. DDT-induced feminization of gull embryos. Science (Wash. D.C.) 213:922–924.

FRY, D. M., C. K. TOONE, S. M. SPEICH, AND R. J. PEARD. 1987. Sex ratio skew and breeding patterns of gulls: demographic and toxicological considerations. Studies in Avian Biology 10:26–43.

FURNESS, R. W., AND T. R. BIRKHEAD. 1984. Seabird colony distributions suggest competition for food supplies during the breeding season. Nature 311:655–656.

HAND, J. L. 1980. Nesting success of Western Gulls on Bird Rock, Santa Catalina Island, California. Pp. 467–473 *in* D. M. Power (ed.). The California Islands: Proceedings of a multidisciplinary symposium. Santa Barbara Mus. of Nat. History, Santa Barbara, Calif.

HAND, J. L. 1986. Territory defense and associated vocalizations of Western Gulls. J. Field Ornithol. 57:1–15.

HUNT, G. L., JR., AND M. W. HUNT. 1975. Reproductive ecology of the Western Gull: the importance of nest spacing. Auk 92:270–279.

HUNT, G. L., JR., R. PITMAN, M. NAUGHTON, K. WINNETT, A. NEWMAN, P. KELLY, AND K. BRIGGS. 1979. Summary of marine mammal and seabird surveys of the southern California bight area 1975–1978, vol. 3, Inv. Rept. 3 to Bur. Land Manag., Univ. of Calif., Irvine.

HUNT, G. L., JR., J. C. WINGFIELD, A. NEWMAN AND D. S. FARNER. 1980. Sex ratio of Western Gulls on Santa Barbara Island, California. Auk 97:473–479.

HUNT, G. L., JR, A. L. NEWMAN, M. H. WARNER, J. C. WINGFIELD, AND J. KAIWI. 1984. Comparative behavior of male-female and female-female pairs among Western Gulls prior to egg-laying. Condor 86:157–162.

KADLEC, J. A., AND W. H. DRURY. 1968. Structure of the New England Herring Gull population. Ecology 49:644–676.

Lack, D. 1954. The natural regulation of animal numbers. Clarendon Press, Oxford.

MCCASKIE, G. 1983. Another look at the Western and Yellow-footed gulls. West. Birds 14:85–107.

MEYERS, J. H. 1978. Sex ratio adjustment under food stress: maximization of quality or numbers of offspring. Am. Nat. 112:381–388.

MONAGHAN, P. 1980. Dominance and dispersal between feeding sites in the Herring Gull, *Larus argentatus*. Anim. Beh. 28:521–527.

PARSONS, J. 1971. The breeding biology of the Herring Gull, *Larus argentatus.* Unpubl. Ph.D. dissertation, Univ. Durham, England.

PIEROTTI, R. 1981. Male and female parental roles in the Western Gull under different environmental conditions. Auk 98:532–549.

PITMAN, R. L., M. R. GRAYBILL, J. HODDER, AND D. H. VAROUJEAN. In press. The catalogue of Oregon Seabird Colonies. U.S. Dept. Int., Fish Wildl. Serv., FWS/OBS.

RICKLEFS, R. E. 1968. Patterns of growth in birds. Ibis 110:419–451.

ROSKAFT, E., AND T. SLAGSVOLD. 1985. Differential mortality of male and female offspring in experimentally manipulated broods of the Rook. J. Anim. Ecol. 54:261–266.

RYDER, J. P. 1983. Sex ratio and egg sequence in Ring-billed Gulls. Auk 100:726–728.

SAYCE, J. R., AND G. L. HUNT, JR. 1987. Sex ratios of prefledging Western Gulls. Auk 104:33–37.

SEARCY, W. A. 1978. Foraging success in three age classes of Glaucous-winged Gulls. Auk 95:586–588.

SOWLS, A. L., A. R. DEGANGE, J. E. NELSON, AND G. S. LESTER. 1980. Catalogue of California Seabird Colonies. U.S. Dept. Inter. Fish Wildl. Serv., FWS/OBS.

SPEAR, L. 1979. A closer look at a neglected resource. Pac. Seabird Gp. Bull. 6(2):33 (abstract).

SPEAR, L. 1980. Band loss from the Western Gull on Southeast Farallon Island. J. Field Ornith. 51:319–328.

SPEAR, L. 1981. Sexing of Farallon Western Gulls by size, and their age/sex related patterns of movement during 1978–1981. Pac. Seabird Gp. Bull. 8(2):96 (abstract).

SPEAR, L. B., D. G. AINLEY, AND R. P. HENDERSON. 1986. Post-fledging parental care in the Western Gull. Condor 88:194–199.

SPEICH, S. M., AND T. R. WAHL. In press. Catalogue of Washington Seabird Colonies. U.S. Dept. Inter. Fish Wildl. Serv., FWS/OBS.

VERBEEK, N. A. M. 1977. Comparative feeding behavior of immature and adult Herring Gulls. Wilson Bull. 89:415–421.

WARD, J. G. 1973. Reproductive success, food supply and the evolution of clutch size in the Glaucous-winged Gull. Unpubl. Ph.D. dissertation, Univ. of British Columbia, Vancouver, British Columbia.

Studies in Avian Biology No. 10:57–65, 1987.

EFFECTS OF INCREASED POPULATION SIZE IN HERRING GULLS ON BREEDING SUCCESS AND OTHER PARAMETERS

ARIE L. SPAANS, ALLE A. N. DE WIT, AND MARIANNE A. VAN VLAARDINGEN[1]

ABSTRACT.—Nest density, clutch initiation, egg volume, reproductive success, and chick growth of Herring Gulls (*Larus argentatus*) were studied in a 6-ha study plot on Terschelling Island, Netherlands, just before and after a dramatic increase in the numbers of breeding pairs on the island (1966–1969: 6000–8000 pairs, 1983–1984: 21,000–21,500 pairs). During the same time interval, Lesser Black-backed Gulls (*L. fuscus*), which form mixed colonies with Herring Gulls, also increased in numbers.

In 1983–1984, almost 100 pairs of large gulls bred per hectare, a 3-fold increase of the nest density compared to 1967–1969. In 1983–1984, Herring Gulls bred, on average, 5–9 days earlier than in 1967–1969 (1967–1969: 18.5–19.4 May, 1983–1984: 10.2–14.0 May). In 1983–1984 reproductive success was 0.34–0.44 fledged young per pair, compared to 1.25–1.50 young per pair in 1967–1969. The lower reproductive success resulted largely from increased predation by conspecifics. Breeding success was highest for early breeders, decreasing in 1983–1984 to almost zero for late breeders. As a consequence, early breeders are at present almost the only ones which contribute genetic material to subsequent generations. We suggest that this situation may have contributed to the shift in clutch initiation that has taken place since the late 1960s. In 1983, the volumes of all eggs in clutches of 3 were smaller than in 1969. In 1984, only b- and c-eggs had smaller volumes. In 1983–1984, chicks gained weight more slowly than in 1966–1968, resulting in a lower fledging weight than earlier. Wing length, however, grew as fast as in 1967–1968, suggesting that chicks were of the same size as in the 1960s but less well-fed than earlier ones.

The results of the study confirm the hypothesis of Coulson et al. (1982) that intraspecific competition is a main factor in the regulation of breeding success and other variables in the population dynamics of the species.

Gulls of the genus *Larus* are highly sociable birds. They breed in colonies and feed mainly in flocks. Such a behavioral pattern implies regular direct confrontations with conspecifics and other seabirds during breeding and feeding. Several studies have shown that gulls interfere with each other on the breeding grounds (e.g., Tinbergen 1960, Brown 1967, Parsons 1971, Hunt and Hunt 1975, Davis and Dunn 1976, Burger 1979, 1984) and in the feeding areas (e.g., Carrick and Murray 1964, Greig et al. 1983, Jones 1985).

Recently, Coulson et al. (1982) hypothesized that for Herring Gulls (*L. argentatus*) several factors associated with the population dynamics of the species—such as nest density, age of recruitment, proportion of young returning to the colony to breed, size of breeding birds and eggs, and reproductive success—are affected by the size of the local population. From 1972 to 1981, Herring Gulls on the Isle of May (Scotland) decreased considerably in numbers as a result of an intense cull program. During the same time, both the density of breeding birds and the age of recruitment decreased, but the proportion of young returning to the colony to breed, body weight and wing length of breeding gulls, and the volume of eggs increased. Coulson et al. (1982) ascribed the changes to reduced intraspecific competition, either on the breeding colony or in the feeding areas, resulting from the decreased population level. If Coulson et al.'s (1982) hypothesis is correct, opposite changes in these parameters may be expected in an increased population. In this paper, we present the results of such a study.

In 1966–1969, just before Herring Gulls started to increase in numbers in The Netherlands, a study of the breeding biology of the species was made in the Boschplaat nature reserve, located on the eastern part of Terschelling, Dutch Frisian Islands (Fig. 1; Spaans and Spaans 1975). Since then the number of Herring Gulls on the island has tripled—from 6000–8000 pairs in 1966–1969 to 21,000–21,500 pairs in 1983–1984 (Fig. 2). During the same time interval, the numbers of Lesser Black-backed Gulls (*L. fuscus*), which form mixed colonies with Herring Gulls, increased from less than 1000 pairs to 13,000 pairs. For the Lessser Black-backed Gull, numbers have been stable since 1982; for the Herring Gull, numbers have decreased since 1985. The considerable increase in gull numbers during the 1970s started 3–4 years after the Herring Gull control program, initiated on the island in the late 1930s (Drost 1939), had been terminated.

STUDY AREA AND METHODS

The 4400-ha Boschplaat Nature Reserve (approximately 53°25′N, 05°28′E) comprises 5 low, undulating dune complexes (Eerste, Tweede, Derde and Vierde

[1] Research Institute for Nature Management, Postbox 9201, 6800 HB Arnhem, Netherlands.

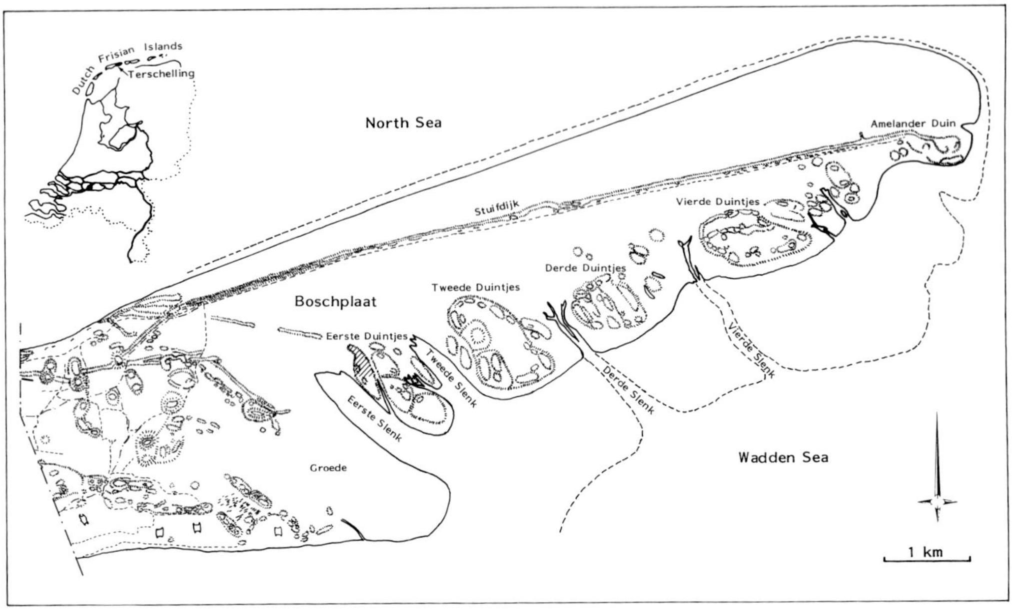

FIGURE 1. The eastern part of Terschelling, showing the Boschplaat Nature Reserve with the study plot (hatched) and the geographic names mentioned in the text.

Duintjes, and Amelander Duin); they are abutted by wide salt marshes and tidal mudflats, which are incised by 4 large tidal creeks (Eerste, Tweede, Derde, and Vierde Slenk) and a number of smaller ones (Fig. 1). The area is protected to the north by a semi-natural sand dam (Stuifdijk) covered by marram grass (*Ammophila arenaria*). The western half of the reserve, up to the Derde Slenk, is closed to the public during the breeding season. The Groede, and the Eerste and Tweede Duintjes are the sites of large and long-established colonies of Herring Gulls and, since 1926, of Lesser Black-backed Gulls also. Only small numbers of gulls breed outside these areas. In 1983–1984, almost 25% of the Dutch Herring Gulls, and more than 60% of the Lesser Black-backed Gulls, bred in the reserve. Between the late 1930s and the middle of the 1960s, Herring Gulls were vigorously controlled by the poisoning of breeding birds and by the collection, oiling, and shaking of eggs. As a result, in the 1960s, the breeding population may have been younger than in the 1980s (cf. Duncan 1978, Coulson et al. 1982).

In 1966, a 6.06-ha study plot was selected on the northwestern part of the Eerste Duintjes (Fig. 1). In choosing a study site, we attempted to select a plot that comprised a large area of the typical breeding habitat of the Herring Gull, was free of disturbance by man and livestock, and could easily be reached without disturbing other breeding birds in the reserve. In both decades, the study plot included parts of the center and edges of the colony. The dunes in the study plot were covered predominantly with marram grass and buckthorn (*Hippophaë rhamnoides*), with local patches of lyme grass (*Leymus arenarius*), woody nightshade (*Solanum dulcamare*), and elder (*Sambucus nigra*). The dunes were interspersed with small grassy valleys dominated by sea couch (*Elymus pycnanthus*). Herring Gulls bred mainly in the dunes, Lesser Black-backed Gulls predominantly in the valleys. The microhabitat of the plot was typical of the whole colony. In the 1980s, the dune complexes were covered with more vegetation than in the 1960s. However, there were fewer burrows of rabbits (*Oryctolagus cuniculus*) than in the 1960s, so that the possibilities for chicks to hide themselves against bad weather and predators did not differ much in the two periods.

In 1966, 150 nests randomly chosen within the study plot at the peak of egg laying were studied until late July. In 1967–1969 and 1983–1984 the study plot was searched, from late April through late June, daily for new clutches in such a way that the entire area was covered at least once every 2 days. In 1967–1969, all clutches found in the study plot were followed up to fledging of the young or to failure. In 1983, all clutches were followed up to hatching or failure, but only one out of every three found was further followed. In 1984, one of every three clutches found was further studied. Incomplete clutches and nests with eggs at hatching were checked once or twice a day to establish the dates of egg laying and hatching as accurately as possible.

Each nest was marked with a small numbered stick as soon as eggs were found. Each egg was marked a, b, or c, according to its laying order. In 1969 and in 1983–1984, maximum length and breadth of the eggs were measured to 0.01 mm using vernier calipers. Egg volume was calculated from: Volume = 0.5035 × Length × Breadth^2 (Spaans and Spaans 1975). In the 1960s data on clutch size and egg volume also included those from a small number of clutches from a small area adjacent to the study plot.

Chicks were individually color-ringed on the day of hatching (day 0). Fledging success was calculated from sightings of fledged young in the immediate vicinity of

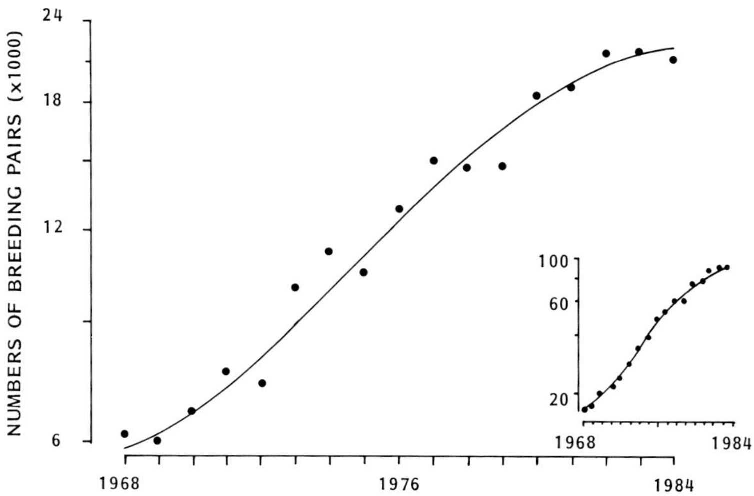

FIGURE 2. The numbers (logarithmic scale) of breeding pairs of Herring Gulls on Terschelling since 1968; the inset shows the population trend for the species in the entire country (curves fitted by eye). The data are taken from the annual Herring Gull censuses organized by the Research Institute for Nature Management and the State Forest Service.

the gullery between early July and late August. These data were supplemented with those of chicks seen alive in the fifth or sixth week of life and not found dead in the colony afterwards nor seen alive just after fledging, and of young that fledged unnoticed (as appeared from sightings or recoveries elsewhere later). The reports of gulls color-ringed at hatching in 1983–1984, when coverage outside the colony area was most intense, indicate that only a very few birds escaped notice just after fledging (Table 1).

Chicks were weighed to the nearest 1.0 g, using a spring balance, on each day they were found alive and healthy. In 1967–1968 and 1984, maximum wing chord was measured by ruler, from day 18, to the nearest 1.0 mm. Between days 4 and 30, the weight of chicks increased approximately linearly. From day 30, it remained approximately the same until fledging (Spaans 1971). Wing length increased linearly from day 18 at least until day 40, which is quite near the average day of fledging. The mean daily weight increases between days 4 and 30, and the mean daily increases in wing length between days 18 and 40 are used here as measures of chick growth. Data are presented according to original brood size and hatching date. Increase in weight and in wing length of cohorts of chicks were estimated by linear regression of weight and wing length on age, allowing for increasing variance with age. Although most chicks were weighed and measured more than once, only one randomly chosen datum per chick was used in order to minimize dependences between observations.

Although the field work was not conducted by the same people in the 1960s as in the 1980s, all worked in a very standardized manner and with about the same levels of contact with the gulls, so that a direct comparison of the results between the two decades seems valid.

Analyses of variance with Poisson-error distribution (McCullagh and Nelder 1983) were used to compare the dates of clutch initiation and the clutch sizes between years and decades. The χ^2-test for two independent samples and the Fisher test were used to test differences in the annual mean hatching/fledging success and in the annual mean percentages of the causes of egg/chick loss between the two decades. The volumes of a-, b-, and c-eggs within the same clutches were compared by means of paired t-tests. Differences in egg volume between years were tested by fitting parallel curves of volume against date of clutch initiation, and comparing intercepts of different years by the t-test. There was some, but not conclusive, evidence of divergence between the curves, but this was mainly a result of the small sample size for clutches intiated between 21 and 26 May 1984. Differences in mean daily increase of weight and wing length between years also were tested with t-tests. All tests are two-tailed.

RESULTS

Nest Density

Between 172 and 201 (mean 185) clutches of Herring Gulls were found in the study plot in 1967–1969, and 526–549 (mean 538) clutches were found in the same area in 1983–1984, ex-

TABLE 1
The Numbers of Herring Gulls Recorded Outside the Colony Area to 1 January 1986 According to Their Fate after Hatching

Fate of chicks	1983	1984
Fledged	23 (78)[2]	20 (60)
Unknown[1]	1 (110)	1 (182)

[1] Chicks which were neither seen alive just after fledging nor found dead during the chick stage.
[2] Numbers in brackets indicate initial numbers of chicks.

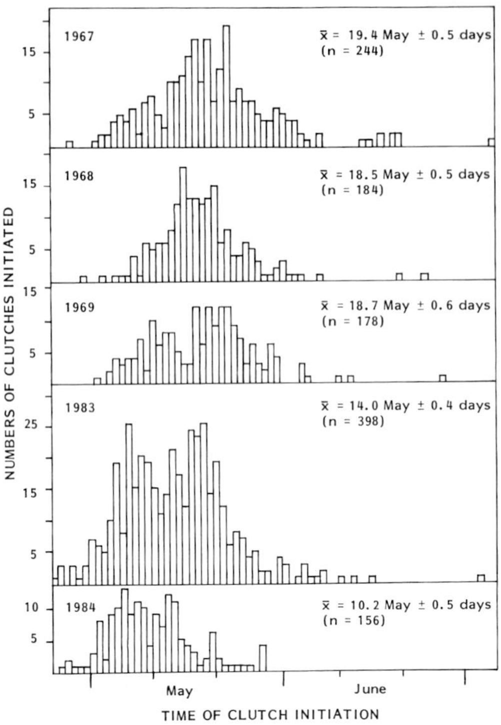

FIGURE 3. Frequency histograms of clutch initiations for 1967–1969 and 1983–1984. Mean dates ± SE for clutches initiated before 11 June (1967–1969) or 6 June (1983–1984) are also given (see the text for reason of choice of different dates).

cluding a very few certain replacements in each year. In addition, between 46 and 50 (mean 48) clutches of Lesser Black-backed Gulls were found in 1983–1984, compared to only one clutch in 1967 (none in 1968–1969). The low numbers of Lesser Black-backed Gulls in the study plot resulted mainly from the very small area of the typical breeding habitat of the species in the plot.

The mean density for the two species of gulls together was 30.5 (28.4–33.2) nests·ha^{-1} in the late 1960s and 96.7 (94.4–98.9) nests·ha^{-1} in the early 1980s. The percentage increase in nest density of the Herring Gull in the study plot was almost as great as that in the total number of the species on Terschelling. This indicates that the growth of the population here has resulted predominantly in a decrease of the internest distances rather than in an extension of the area occupied by the gulls. This was corroborated by our field observations.

Mean Date of Clutch Initiation

In the years for which we have data from the entire egg-laying season, the distribution of the dates of clutch initiation was not always unimodal (Fig. 3). Discrepancies from unimodality could be ascribed to delays in clutch initiation during periods of cold and wet weather. Besides, the distribution was in all years skewed toward the end of the season as a result of late replacements. Based on the distribution pattern over the season, we identified the end of the laying season for first clutches to be 10 June (1967–1969) and 5 June (1983–1984), respectively. For calculation of the mean dates of clutch initiation we used only clutches started at or before these dates.

There is a significant difference between the annual mean dates of clutch initiation (analysis of variance, $\chi^2 = 158.3$, df = 4, $P < 0.001$). In 1967–1969, the mean date of clutch initiation was 18.9 May and in 1983–1984 12.1 May, and were significantly different from one another (analysis of variance, $\chi^2 = 111.9$, df = 1, $P < 0.001$).

Egg Volume

In two of the three years for which we have data on egg volumes, we found a small but significant difference between the mean volumes of a- and b-eggs within the same clutch (Table 3, 1969: t = 0.72, df = 71, $P > 0.20$; 1983: t = 2.98, df = 186, $P < 0.01$; 1984: t = 5.35, df = 188, $P < 0.001$). In all three years, both a- and b-eggs were, on average, much larger than c-eggs (1969, a–c: t = 10.29, df = 71, $P < 0.001$; b–c: t = 11.90, df = 71, $P < 0.001$; 1983, a–c: t = 18.16, df = 186, $P < 0.001$; b–c: t = 18.17, df = 186, $P < 0.001$; 1984, a–c: t = 20.90, df = 188, $P < 0.001$; b–c: t = 20.96, df = 188, $P < 0.001$).

In 1983, all eggs in clutches of three with at least one egg of which the laying order was known were significantly smaller than their counterparts in 1969, the difference being most pronounced

TABLE 2
Mean Volume (ml) ± s.e. of First (a), Second (b), and Third (c) Laid Eggs of Herring Gulls on Terschelling in 1969 and 1983–1984

Year	No. of clutches	a-egg	b-egg	c-egg
1969	73	89.3 ± 0.71	89.0 ± 0.70	83.5 ± 0.71
1983	188	87.1 ± 0.51	86.1 ± 0.47	80.2 ± 0.47
1984	190	90.1 ± 0.51	88.3 ± 0.43	81.9 ± 0.41

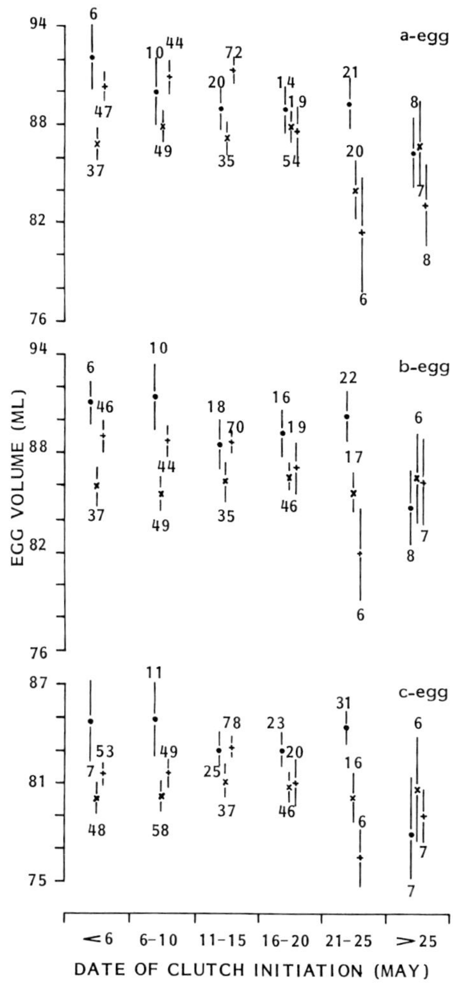

FIGURE 4. Annual mean egg volume (ml) ± SE of first (a), second (b), and third (c) laid eggs of Herring Gulls on Terschelling before (1969, dots) and after (1983–1984, crosses, plus signs, respectively) the numerical increase in breeding pairs according to the date of clutch initiation. Numbers given are sample sizes.

in the b- and c-eggs (Fig. 4, a-eggs: t = −2.75, df = 472, $P < 0.01$; b-eggs: t = −4.50, df = 457, $P < 0.001$; c-eggs: t = −4.60, df = 523, $P < 0.001$). In 1984, the difference was significant only in the c-eggs, but nearly so in the b-eggs (a-eggs: t = 0.07, df = 472, $P > 0.20$; b-eggs: t = −1.91, df = 457, $P < 0.10$; c-eggs: t = −2.62, df = 523, $P < 0.01$). As can be seen from Figure 4, the smaller egg volumes hold for both early and late breeders.

Reproductive Success

The mean clutch sizes for 1967–1969 and 1983–1984 are shown in Table 3. Two of the

TABLE 3
CLUTCH SIZE OF HERRING GULLS ON TERSCHELLING BEFORE (1967–1969) AND AFTER (1983–1984) THE NUMERICAL INCREASE IN BREEDING PAIRS

Year	Clutch size 4	3	2	1	Mean
1967	0	180	47	14	2.69
1968	0	147	30	5	2.78
1969	0	138	33	5	2.76
1983	1	331	155	62	2.49
1984	1	133	25	15	2.69

1322 clutches had 4 eggs. In 1983, 2 females were involved. The same may have been the case for the c/4 in 1984. For this reason, the 2 c/4 are further left out of consideration. Mean clutch sizes differed significantly between the five years (analysis of variance, $\chi^2 = 56.9$, df = 8, $P < 0.001$). In 1967–1969, the mean clutch size was 2.74 and in 1983–1984 2.54. These were significantly different from one another (analysis of variance, $\chi^2 = 35.0$, df = 2, $P < 0.001$). The difference between the two periods may be an artefact. Not all eggs were found immediately after laying. As a result of the larger predation pressure in the 1980s compared to the 1960s (Tables 4, 5), relatively more eggs may have been robbed unnoticed in the 1980s than earlier.

Hatching and fledging success were quite different in the two periods. In the 1980s, mean hatching success was 61.5%, which was much lower than that in the 1960s (76.5%, Table 4). The difference was due largely to an increase in the percentages of eggs eaten by conspecifics (Table 4). The percentages of addled eggs and other causes of egg loss did not change. There was also a large reduction in the percentage of hatched chicks that fledged (Table 5). In the 1960s, mean fledging success was 58.2%, in the 1980s 23.2%. Predation by conspecifics, which may include some scavenging, was again the main cause of the losses (Table 5). The distribution of regurgitated color-rings over the colony suggested that predation was mainly by non-neighbors rather than by gulls from adjacent territories.

The reduction in hatching and fledging success held for all gulls, regardless of the date of clutch initiation (Figs. 5, 6). As a result of the large numbers of eggs and chicks lost in the 1980s, the number of fledged young per pair decreased from 1.25–1.50 (mean 1.35) in 1967–1969 to 0.34–0.44 (mean 0.39) in 1983–1984, a reduction of more than 70%, on average. At present, predation increases so much during the course of the breeding season, that even birds laying around the mean date of clutch initiation produce hardly any young to the age of fledging.

TABLE 4

COMPARISON OF ANNUAL MEAN HATCHING SUCCESS AND CAUSES OF EGG LOSS BY PERCENTAGES (MEANS, RANGES) OF EGGS LAID BY HERRING GULLS ON TERSCHELLING BEFORE (1967–1969) AND AFTER (1983–1984) THE NUMERICAL INCREASE IN BREEDING PAIRS

Fate of eggs	1967–1969	1983–1984	Tests of significance of differences
Hatched	76.5 (75.5–77.9)	61.5 (54.8–68.1)	$\chi^2 = 78.9, P < 0.001$
Predation[1]	8.2 (6.8–9.9)	22.4 (15.0–29.8)	$\chi^2 = 114.0, P < 0.001$
Infertility/embryonic death	13.2 (11.7–15.1)	14.4 (14.0–14.8)	$\chi^2 = 0.9$, n.s.[2]
Died during hatching	1.2 (1.0–1.5)	1.6 (0.4–2.8)	$\chi^2 = 0.9$, n.s.
Rolled out of nest	0.8 (0.4–1.3)	0.0 (0.0)	Fisher test, n.s.
Destroyed by observer	0.1 (0.0–0.2)	0.1 (0.0–0.1)	Fisher test, n.s.
No. of eggs studied	1374	1697	

[1] Largely by conspecifics.
[2] n.s. = not significant.

SIZE OF FLEDGLINGS

In 1983–1984 chicks gained weight slower than in 1966–1968, in 17 of the 22 possible pairwise comparisons between the two periods (Fig. 7). In two cases, the difference was significant (b/3, 12–19 May, 1966–1983: t = 2.41, df = 33, $P < 0.05$; b/3, <6 May, 1967–1984: t = 2.74, df = 37, $P < 0.01$). The slower weight increase resulted in a lower weight of the chicks at day 30 than in the 1960s, as calculated from the regression equations. In five cases, the difference was significant (b/3, <6 May, 1967–1984: t = 4.10, df = 37, $P < 0.001$; b/3, 6–11 May, 1968–1983: t = 2.65, df = 13, $P = 0.02$; b/3, 12–19 May, 1966–1983: t = 2.90, df = 33, $P < 0.01$; b/3, 12–19 May, 1967–1983: t = 2.59, df = 46, $P < 0.02$; b/2, 6–11 May, 1967–1984: t = 2.81, df = 6, $P < 0.05$), and in two cases nearly so (b/3, 6–11 May, 1967–1983: t = 1.91, df = 14, $P < 0.10$; b/2, 12–19 May, 1967–1983: t = 3.09, df = 3, $P < 0.10$). In both decades, weight varied around the 30th day level at least until fledging, so that the differences in weight at day 30 must have been reflected in the weight of the birds at fledging.

We have fewer data on wing length than on weight. However the few data we have suggest that in the 1980s, wing length between days 18 and 40 increased as fast as in the late 1960s (Table 6, b/3, <6 May, 1967–1984: t = −0.82, df = 19, $P > 0.20$; b/3, <6 May, 1968–1984: t = −0.05, df = 14, $P > 0.20$; b/3, 6–11 May, 1967–1984: t = −1.43, df = 15, $P > 0.20$; b/3, 6–11 May, 1968–1984: t = −1.62, df = 15, $P > 0.10$; b/3, 12–19 May, 1967–1984: t = 0.46, df = 26, $P > 0.20$; b/3, 12–19 May, 1968–1984: t = 0.72, df = 3, $P > 0.20$). In none of the cases did wing length on day 40 differ significantly ($P > 0.20$). These results suggest that in the 1980s young were of the same size as in the 1960s but less well-fed than earlier.

DISCUSSION

The total number of large gulls on Terschelling has shown a 5-fold increase over the last 15 years. The larger population has resulted in a 3-fold increase of the nest density within our study plot, where in 1983–1984 almost 100 pairs of large gulls bred per hectare.

TABLE 5

COMPARISON OF ANNUAL MEAN FLEDGING SUCCESS AND CAUSES OF CHICK LOSS BY PERCENTAGES (MEANS, RANGES) OF HERRING GULL CHICKS ON TERSCHELLING BEFORE (1967–1969) AND AFTER (1983–1984) THE NUMERICAL INCREASE IN BREEDING PAIRS

Fate of chicks	1967–1969	1983–1984	Tests of significance of differences
Fledged	58.2 (56.4–60.4)	23.2 (19.1–27.2)	$\chi^2 = 175.8, P < 0.001$
Predation/scavenging	9.8 (5.8–13.6)[1]	23.0 (21.3–24.7)	$\chi^2 = 162.9, P < 0.001$
Other causes of death		14.7 (13.2–16.2)	
Fate unknown[2]	32.0 (28.2–33.8)	39.2 (38.3–40.0)	$\chi^2 = 7.8, P < 0.01$
No. of chicks studied	865	596	

[1] No distinction made between predation/scavenging and other causes of death, because only a very few chicks fell in the former category in those years.
[2] Almost all these chicks died unnoticed before fledging (see Table 1).

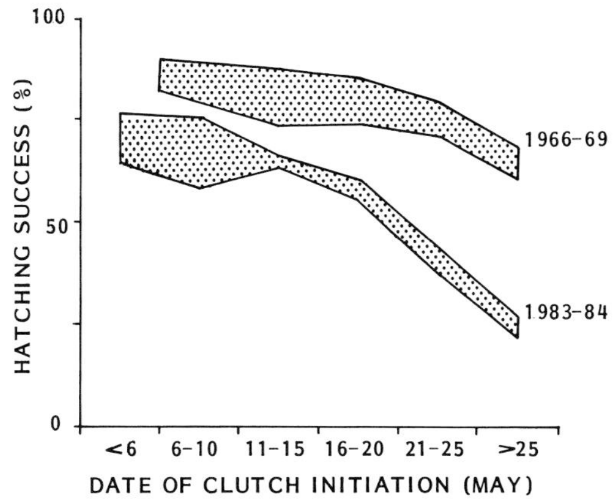

FIGURE 5. Annual mean hatching success (percentages, ranges) of Herring Gulls on Terschelling before (1966–1969) and after (1983–1984) the numerical increase in breeding pairs according to the date of clutch initiation.

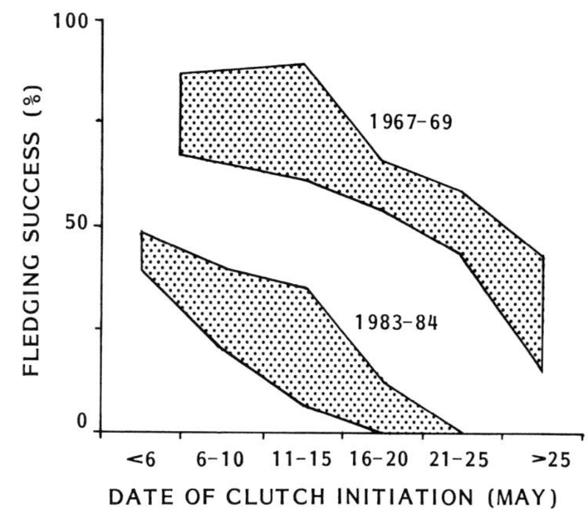

FIGURE 6. Annual mean fledging success (percentages, ranges) of Herring Gulls on Terschelling before (1967–1969) and after (1983–1984) the numerical increase in breeding pairs according to the date of clutch initiation.

We found marked changes in the breeding biology of the Herring Gull in the 1980s compared to the late 1960s. The number of fledged young per pair in the 1980s decreased by more than 70%, largely resulting from an increase in the percentages of eggs and chicks robbed by conspecifics. Moreover, eggs decreased in size, and chicks gained weight more slowly and fledged at lower weight. In fact, the differences in growth between the two decades must have been even larger than those shown, for two reasons. Firstly, in the 1980s, as a result of the higher chick mortality, many more original b/3 and b/2 decreased in size than in the 1960s. Most of the mortality occurred in the first week after hatching. As a consequence, in the 1980s chicks met less competition for food after the first few days of life than in the 1960s and should therefore have grown better than earlier. Secondly, c-chicks are smaller than a- and b-chicks and hatch a few days after these, resulting in much higher post-hatching mortality of the c-chicks (Parsons 1970, Spaans and Spaans 1975). Therefore, in the 1980s, relatively more a- and b-chicks, which are not only larger but also grow faster than c-chicks, remained alive than in the 1960s. This factor should also have positively influenced mean weights in the 1980s. Yet, in these years, chicks

TABLE 6

MEAN DAILY INCREASE PER ANNUM IN WING LENGTH (MM) BETWEEN DAYS 18 AND 40 ± S.E. OF HERRING GULL CHICKS ON TERSCHELLING BEFORE (1967–1968) AND AFTER (1984) THE NUMERICAL INCREASE IN BREEDING NUMBERS[1]

Brood size	Date of clutch initiation (May)	1967	1968	1984
b/3	<6	7.0 ± 3.8 (8)	9.0 ± 0.6 (3)	10.1 ± 0.8 (15)
b/3	6–11	8.6 ± 0.6 (8)	8.1 ± 0.5 (8)	10.4 ± 1.1 (11)
b/3	12–19	9.0 ± 0.5 (26)	9.1 ± 0.5 (3)	8.2 ± 0.5 (4)
b/3	20–26	6.8 ± 1.3 (12)	9.4 ± 0.9 (16)	
b/3	>26	12.4 ± 2.7 (4)		
b/2	<6			6.3 ± 2.9 (5)
b/2	6–11			8.9 ± 1.5 (5)
b/2	12–19	9.3 ± 2.2 (3)		
b/2	20–26	8.0 ± 3.7 (3)		
b/1	6–11	7.4 ± 0.9 (5)	8.1 ± 0.1 (3)	
b/1	12–19	8.9 ± 0.6 (10)		
b/1	20–26	7.6 ± 0.8 (9)	10.9 ± 2.0 (7)	
b/1	>26	9.7 ± 1.2 (5)		

[1] Data are given according to the original brood size and the date of clutch initiation. Numbers of chicks measured indicated in brackets.

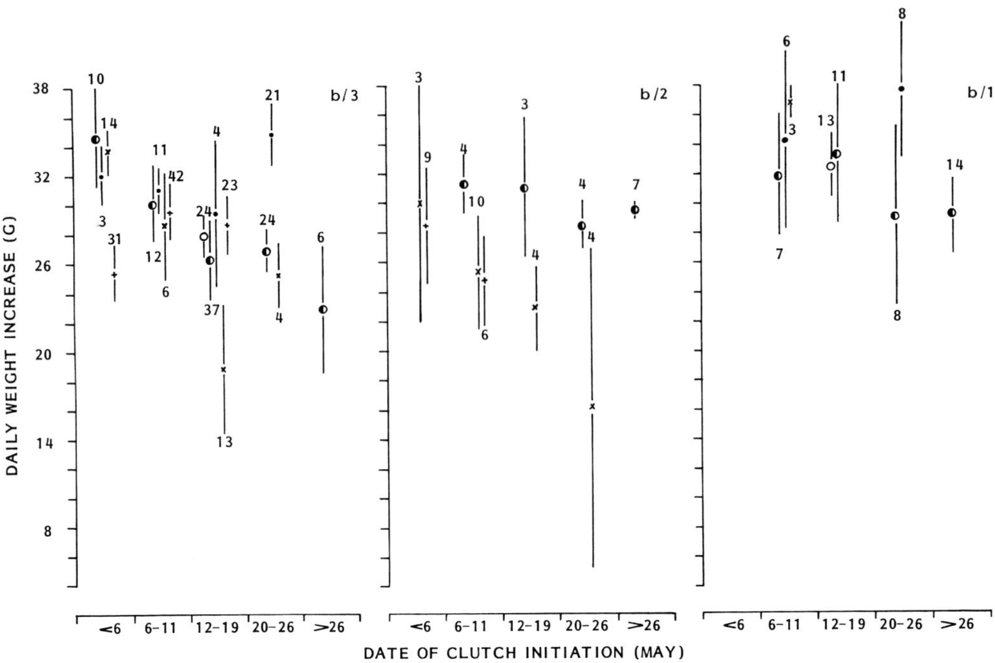

FIGURE 7. Mean daily weight increase per annum between days 4 and 30 (g) ± SE of Herring Gull chicks on Terschelling before (1966–1968, open, half-open, and black dots, respectively) and after (1983–1984, crosses, plus signs, respectively) the numerical increase in breeding pairs according to the original brood size and the date of clutch initiation. Numbers given are sample sizes.

gained weight more slowly and reached lower fledging weights than did earlier ones.

Considering Figures 4–7, it can be argued that it would be more appropriate to compare egg volume, breeding success, and chick growth in relation to the peak of egg-laying rather than to the date of clutch initiation. We did this and found that the trends remain the same. Therefore, we conclude that the negative changes we found are due not only to a shift in the timing of breeding, but also to some other factor(s).

One may hypothesize that any increase in predation rates may have resulted from observer disturbance of pairs at higher densities. We reject this idea because the low breeding success was not limited to the study plot but was characteristic of the entire colony, as shown from the ratio adults/fledged young at the colony edge during low tide at the end of the breeding season (more young than adults in the late 1960s and the opposite in the 1980s). Several aspects of the breeding performance of Herring Gulls such as date of clutch initiation and egg size are known to vary with age, females laying progressively earlier up to at least 9 years old and laying eggs progressively increasing in volume up to 7 or 8 years of age and then declining (Davis 1975). As a result of the intense cull program conducted on the island between the late 1930s and the middle of the 1960s, the Herring Gull population may have been younger at the beginning of the study period than at the end. The decrease in the volume of eggs on the island, however, held for all cohorts irrespective of the date of clutch initiation (Fig. 4). Such a pattern is unlikely if only an age effect were operative. An age effect, however, cannot be excluded as a reason for the shift that took place in the mean date of clutch initiation since the 1960s (cf. Davis 1975). Alternatively, the strong positive selection—resulting from the seasonal trend in egg and chick predation—in favor of gulls which lay their eggs early in the season may have played a role.

Coulson et al. (1982) showed that several factors of population dynamics changed in a colony that had been reduced to a quarter of its former level. Thus, an appreciable increase in egg size, wing length, and body weight of the breeding birds occurred, while there was also some evidence for an improvement of the breeding success. Moreover, there was an increase in the proportion of third- and fourth-year birds in the breeding population. We have no data on the age at which recruits enter the population at Terschelling, but in 1983–1984, breeding birds retaining feathers typical of immature gulls were seen by us occasionally. The same pattern was observed by Coulson et al. (1982) prior to the starting of the cull program on the Isle of May; it is interesting to note that nest density at that time was approximately the same as was the density on Terschelling in 1983–1984.

The changes found by us in a population that had increased are, therefore, just opposite to those found by Coulson et al. (1982) in a population that had decreased, confirming the prediction put forward in the introduction. Herring Gulls clearly show marked changes in factors related to the population dynamics of the species when the number of breeding birds fluctuates; the direction of the changes is opposite to the direction of the population trend. However, the question still remains as to whether competition takes place primarily on the breeding grounds or in the feeding areas. The pattern of predation on Terschelling (by non-neighbors rather than by gulls from adjacent territories) suggests that the increase in predation is related there to food shortage rather than to stress on the breeding area (contra de Wit and Spaans 1984). The smaller eggs and the poorer condition of the chicks support the idea of a food shortage (cf. Verbeek 1986).

ACKNOWLEDGMENTS

We gratefully thank the State Forest Service for giving us the opportunity to conduct this study in the Boschplaat Nature Reserve, Harry Horn and his staff for their logistic assistance, Ruurd Noordhuis, Rob L. Vogel and Jan N. de Wit for their help in the data collection, and Remco J. van Rijn for his assistance in carrying out the statistical analyses. W. A. Montevecchi, K. Vermeer, and an anonymous referee provided helpful comments on an earlier version of the manuscript.

LITERATURE CITED

BROWN, R. G. B. 1967. Breeding success and population growth in a colony of Herring and Lesser Black-backed Gulls *Larus argentatus* and *L. fucus.* Ibis 109: 502–515.

BURGER, J. 1979. Competition and predation: Herring Gulls versus Laughing Gulls. Condor 81:269–277.

BURGER, J. 1984. Pattern, mechanism, and adaptive significance of territoriality in Herring Gulls (*Larus argentatus*). Ornithol. Monogr., No. 34.

CARRICK, R., AND M. D. MURRAY. 1964. Social factors in population regulation of the Silver Gull, *Larus novaehollandiae* Stephens. CSIRO Wildl. Res. 9:188–199.

COULSON, J. C., N. DUNCAN, AND C. THOMAS. 1982. Changes in the breeding biology of the Herring Gull (*Larus argentatus*) induced by reduction in the size and density of the colony. J. Anim. Ecol. 51:739–756.

DAVIS, J. W. F. 1975. Age, egg-size and breeding success in the Herring Gull *Larus argentatus.* Ibis 117:460–473.

DAVIS, J. W. F., AND E. K. DUNN. 1976. Intraspecific predation and colonial breeding in Lesser Black-backed Gulls *Larus fuscus.* Ibis 118:65–67.

DE WIT, A. A. N., AND A. L. SPAANS. 1984. Veranderingen in de broedbiologie van de Zilvermeeuw *Larus argentatus* door toegenomen aantallen. Limosa 57:87–90.

DROST, R. 1939. Massnahmen zur Bekämpfung der Silbermöwen in Holland. Naturschutz 21(2):1–3.

DUNCAN, N. 1978. The effects of culling Herring Gulls (*Larus argentatus*) on recruitment and population dynamics. J. Appl. Ecol. 15:697–713.

GREIG, S. A., J. C. COULSON, AND P. MONAGHAN. 1983. Age-related differences in foraging success in the Herring Gull (*Larus argentatus*). Anim. Behav. 31: 1237–1243.

HUNT, G. L., AND M. W. HUNT. 1975. Reproductive ecology of the Western Gull: the importance of nest-spacing. Auk 92:270–279.

JONES, M. J. 1985. Distribution and activity of Common Gulls and Black-headed Gulls on grassland in Manchester. Bird Study 32:104–112.

MCCULLAGH, P., AND J. A. NELDER. 1983. Generalized linear models. Chapman and Hall, London.

PARSONS, J. 1970. Relationship between egg size and post-hatching chick mortality in the Herring Gull (*Larus argentatus*). Nature 228:1221–1222.

PARSONS, J. 1971. Cannibalism in Herring Gulls. Br. Birds 64:528–537.

SPAANS, A. L. 1971. On the feeding ecology of the Herring Gull *Larus argentatus* Pont. in the northern part of The Netherlands. Ardea 53:73–188.

SPAANS, M. J., AND A. L. SPAANS. 1975. Enkele gegevens over de broedbiologie van de Zilvermeeuw *Larus argentatus* op Terschelling. Limosa 48:1–39.

TINBERGEN, N. 1960. The Herring Gull's world. Collins, London.

VERBEEK, N. A. M. 1986. Aspects of the breeding biology of an expanded population of Glaucous-winged Gulls in British Columbia. J. Field Ornithol. 57:22–33.

Studies in Avian Biology No. 10:66, 1987.

SELECTIVE FACTORS AFFECTING CLUTCH SIZE IN THE WESTERN GULL ON THE FARALLON ISLANDS, CALIFORNIA

MALCOLM C. COULTER[1]

ABSTRACT.—Western Gulls, like most gulls, usually lay clutches of three eggs. I examined selective factors affecting clutch size during both the egg and chick stages. I moved eggs and chicks among nests to create clutches and broods that were artificially larger and smaller than the usual three. Western Gulls are highly adapted to care for three eggs. They have three brood patches. They tended to sit for longer spells on three eggs than on larger or smaller clutches, and the incubation period was shortest for three-egg clutches. Hatching success was highest for three-egg clutches. During the chick stage, parents could raise more than three chicks. The greatest number of chicks fledged from five-chick broods. Chicks from the larger broods weighed less than chicks from the smaller broods. When the results from the egg and chick stages are combined, the greatest mortality occurred during the egg stage. I suggest that, at some time in the past, Western Gulls evolved three-egg clutches, and that the gulls have since become highly adapted to care for three eggs. These adaptations presently exert selection pressure maintaining the three-egg clutches.

[1] Savannah River Ecology Laboratory, P.O. Drawer E, Aiken, South Carolina 29802.

Studies in Avian Biology No. 10:67, 1987.

A COMPARISON OF SOME ADAPTATIONS OF HERRING AND RING-BILLED GULL CHICKS TO THEIR NATAL ENVIRONMENT

LINDA M. UIN[1]

ABSTRACT.—I studied aspects of the behavior of chicks of ground-nesting Herring Gulls (*Larus argentatus*) and Ring-billed Gulls (*L. delawarensis*) in the interlake region of Manitoba in 1981 and 1982 to identify possible adaptations of the chicks to their natal environment. Results obtained from both colony- and laboratory-reared chicks suggest that the characteristic mobility pattern exhibited by chicks of each species represents an important adaptation to the chick's particular natal environment. The tendency for Ring-billed Gull chicks to remain sedentary in a cohesive family group at all locations, except to approach and follow their calling parents during movements away from the territory, may represent a strategy designed for areas with high nest densities. Conversely, the lack of parentally induced wandering and the avoidance of hostile neighboring adults, young, and other rearing areas in Herring Gull chicks may represent an adaptation to low nest densities and high aggression levels characteristic of this species.

[1] Department of Zoology, University of Manitoba, Winnipeg, Manitoba R3T 2N2, Canada.

Studies in Avian Biology No. 10:68–74, 1987.

TIME-PARTITIONING OF CLUTCH AND BROOD CARE ACTIVITIES IN HERRING GULLS: A MEASURE OF PARENTAL QUALITY?

RALPH D. MORRIS[1]

ABSTRACT.—Thirty-one Herring Gull (*Larus argentatus*) pairs were observed during incubation and brood care over three breeding seasons at a colony near Port Colborne, Ontario (42°53′N, 79°16′W). Parents that successfully raised two or more chicks (n = 17) were normally both present with the clutch during incubation. In most of these pairs, timing of incubation was partitioned such that each partner incubated most frequently at predictable times of the day. Similar attendance synchrony was recorded during the first 10 days of brood care. In other successful pairs, incubation and brood care were also partitioned equitably between partners. Conversely, in less successful pairs that raised at most one chick (n = 14), clutches and broods were frequently unattended by one or both parents. Synchronous or equitable partitioning of parental care activities were absent, with consequent egg and chick loss. Differential parental quality of pairs is inferred from these patterns.

Gull species (Laridae) are generally monogamous and both partners exhibit extensive parental care behavior during incubation and chick care periods. Among large-bodied *Larus* gulls, joint parental care is particularly important as egg and chick loss to conspecific neighbors can be substantial (Brown 1967, Parsons 1971). Thus, active participation by both parents may be necessary for chicks to be successfully raised.

Studies of several colonially nesting seabirds have revealed qualitative and quantitative sexual differences in parental activities during breeding (Burger 1981, Butler and Janes-Butler 1983). For example, Southern (1981) reported that male Ring-billed Gulls (*L. delawarensis*) were more agonistic toward intruders than females, an observation also noted for male Herring Gulls (*L. argentatus,* Morris and Bidochka 1982). Male Western Gulls (*L. occidentalis*) were more active in territorial defense (Hand 1986) and fed chicks more frequently than their female partners (Pierotti 1981). The approach used in all these studies was to pool data from all pairs to establish sexual differences during breeding.

My intention was to take an alternative approach. I observed the behavior of Herring Gull partners during incubation and brood care periods, and then examined parental contributions on a pair by pair basis. Specific objectives were (1) to record attendance patterns by partners during incubation and the first 10 days of chick age, and (2) to assess the relationship between patterns observed and the reproductive success of each pair.

METHODS

The Herring Gull colony was on an artificial breakwall 1 km off the north shore of Lake Erie near Port Colborne, Ontario (see Morris and Haymes 1977 for descriptive and photographic details). In each of 3 years, observations from an elevated blind, located on the periphery of the colony that numbered between 75–80 pairs, began in mid-April, prior to the initiation of the earliest clutch. The rock substrate is normally free of ice by early March, and although many pairs are present then, egg-laying by the earliest pairs began in mid-April in two of the three years (see Morris and Chardine 1985 for details of the exception).

I observed incubation and chick care attendance by partners of 8 pairs in 1979, 12 pairs in 1980 and 11 pairs in 1981. Study pairs were selected using the following criteria: (1) some identifying mark (aluminum ring, color band) on at least one member of each pair, (2) proximity to the observation blind (within a 15-m radius), (3) first eggs laid during a "peak" of egg laying at the colony, and (4) a completed clutch size of 3 eggs. Some individuals or pairs were observed in more than one year (see Results). Peak egg laying was arbitrarily defined as within ±1 SD of the mean date of first eggs laid by all pairs in the colony that year (cf. Morris and Chardine 1985). Clutches were inspected periodically during incubation and hatching by walking to all nests. A second observer was in the blind during such checks to note the behavior of adults during and following investigator activity in the colony. Chick survival in study broods was determined by observation from the blind.

Parental attendance data were taken almost daily each year during 3-h periods following sunrise or preceding sunset. These periods are known to be times of maximal activity by adult Herring Gulls at this location (Morris and Black 1980). For all single observation days, morning and evening observation periods were rotated such that the total numbers of hours in morning and evening time slots were approximately equal each year. At the beginning of most periods, two people entered the blind and one left immediately; behavior of the gulls was normal within 5 min of departure by the second person. Study pairs were scanned continuously through horizontal openings in the front and two sides of the blind taking care not to focus on any pair for more than a few seconds. Data on parental attendance were recorded directly onto a field sheet formatted to receive information for each pair. Records of attendance with their clutch or brood by members of each pair were recorded from laying of the first egg

[1] Department of Biological Sciences, Brock University, St. Catharines, Ontario L2S 3A1, Canada.

to the date when the youngest chick in the brood was 10 days old. During incubation, I recorded the sex of each incubating bird and the presence or absence of a standing mate. The sex of partners was based on courtship feeding and copulation activities observed early in each year. The presence or absence of the mate not incubating or attending chicks was determined with relative ease as off-duty birds normally occupied a characteristic location near the nest. The number of eggs hatching in each study clutch, and the number of chicks surviving to at least 20 days of age were determined each year.

RESULTS

Parental Reproductive Success

The 3-egg clutches of all study pairs were initiated between 28 April–14 May 1979, 20 April–2 May 1980, and 27 April–3 May 1981. In each year, these dates fell within 1 SD of the mean date of clutch initiation for all pairs in the colony. Based on reproductive performance, pairs were arbitrarily segregated each year into two relative categories of parental reproductive success (Table 1). "Successful" parents (n = 17) were taken as those that hatched 2–3 eggs, and had a minimum of 2 chicks survive to 20 days of age. Less successful parents (n = 14) were those that hatched 0–3 eggs, and had a maximum of 1 chick survive to 20 days of age. Successful and less successful pairs were distributed throughout the range of clutch initiation dates in each year. There was no tendency for either group of parents to initiate clutches before or after the mean date of egg laying (Mann Whitney U test, $n_1 = 14$, $n_2 = 17$, $U = 82$, $P > 0.1$).

Descriptive statistics (percentages, means ± 1 SD) for pairs in the successful and less successful parental groups were as follows. Forty-nine of 51 eggs (96%) hatched in clutches of successful parents, and 42 chicks (85%) survived to at least 20 days from these broods (2.47 ± 0.49 chicks per pair). None of these broods lost more than a single chick. Conversely, of 42 eggs laid by females of less successful pairs, 23 (54%) hatched, and single chicks survived to 20 days from only 8 broods (34%; 0.57 ± 0.48 chicks per pair). Three pairs abandoned their clutches about one week before eggs were due to hatch. Where known, the fates of eggs that did not hatch and the age of chicks that were lost are listed in Table 1.

TABLE 1
Herring Gull Pairs Grouped According to Hatching Success and Chick Survival to 20 Days of Age[a]

Year	Parental pair category[b]: Successful (Hatched 2–3 eggs,[c] 2,3 chicks survived)	Less successful (Hatched 1–3 eggs, 1 chick survived)	Less successful (Hatched 0–2 eggs, no chicks survived)
1979	4	1	3[d,e]
1980	6	3	3[f]
1981	7	4	0

[a] 16 chicks lost between 1–10 days of age; 6 lost between 11–20 days.
[b] See text for average number of chicks fledged by pairs in each parental pair category.
[c] 15 of 17 pairs hatched all 3 eggs.
[d] All eggs in one clutch destroyed by male (Chardine and Morris 1983).
[e] Two clutches abandoned by parents late in incubation.
[f] One clutch abandoned by parents late in incubation.

Ten of the 23 pairs under observation in 1980 and 1981 were known to have changed ("change," 7 pairs) or retained ("same," 3 pairs) their mates from the previous breeding season. Previous-partner status of the other 13 pairs observed in 1980 and 1981, and of the 8 pairs observed in 1979, was unknown. Various combinations of successful and less successful status (as defined above) were achieved by "change" and "same" pairs in the second year. Of the 7 "change" pairs, 5 were successful in both years, 1 was less successful in both years, and 1 was less successful in the second year. Four birds known to have changed partners were male and 3 were female. Of the 3 "same" pairs, 1 was successful in both years, 1 was less successful in both years, and 1 achieved successful status in the second year. Thus, 6 of 7 "change" pairs achieved the same reproductive success status in the two years, while 2 of 3 "same" pairs did so. There was no difference in the distribution of pairs within these two categories of reproductive success (Fisher Exact Probability Test, $P = 0.94$).

Parental Incubation

Incubation time by each partner in each pair was determined as time spent incubating when the partner was gone, plus time spent incubating when both members of the pair were with the clutch (one incubating, one standing). The sum of these components during each hour of observation was used to determine the mean number of minutes per hour each partner spent incubating during morning and evening observation periods. During incubation, the mean observation time allocated to each of the 31 pairs was 48.9 ± 11.8 h.

Ten of the 17 pairs of successful parents exhibited an incubation pattern where one member incubated primarily in the morning while the partner incubated primarily in the evening (Fig. 1A). Incubation was either solitary with the mate gone, or more usually with the mate standing nearby. This "trade-off" pattern in the timing of incubation effort by mates was more strongly marked among some of the ten pairs, but was consistent among them all. There was no tendency for a particular sex to incubate in one time

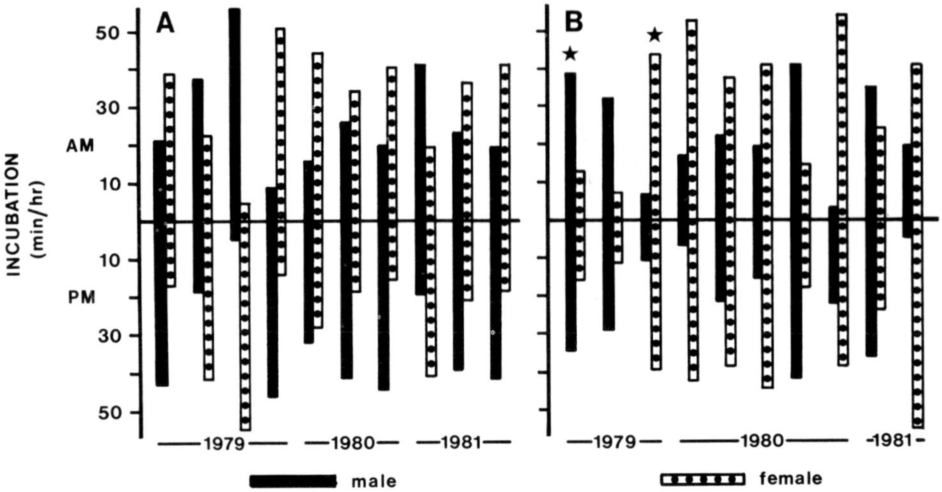

FIGURE 1. Time (mean minutes/hour) spent in incubation of 3-egg clutches by partners of Herring Gull pairs. A. Successful pairs, fledged 2–3 chicks. B. Less successful pairs, fledged 0–1 chick. (See text and Table 1 for further details.) The stars note pairs that abandoned their clutch late in incubation.

period or the other. In the other 7 successful pairs, one partner contributed more time to incubation than its mate in both morning and evening periods. In 4 pairs, the female contributed more to incubation duties, whereas in 3 pairs the male did so. However, the contribution to incubation by the partners was more equitable than disproportionate; the average incubation time each hour by the most committed member of the pair was only slightly more than 30 min (4 females, morning = 35.9 ± 4.1 min/h, evening = 36.9 ± 6.3 min/h; 3 males, morning = 34.2 ± 1.9 min/h, evening = 35.6 ± 1.3 min/h).

Contrary to the incubation behavior of successful parents, the timing of incubation by partners in 10 of the 14 less successful pairs revealed a quite different pattern (Fig. 1B). In each case, one parent incubated much more frequently than the other in both morning and evening periods, and there was no evidence of the incubation "trade-offs" observed between partners in many successful pairs. Of the remaining 4 pairs of less successful parents, data in 2 cases were too few (<10 hours each) to permit assessment; in the other 2 cases, partners did not follow any discernable pattern of incubation. Examples of the two principal types of incubation pattern used by successful (synchronous partitioning) and less successful (one partner heavily committed) pairs were observed in all three years (Fig. 1).

Finally, during our hours of observation, partners of 15 successful pairs were never simultaneously absent from their clutches. Partners from the other 2 pairs were both absent for less than 30 min in the over 40 h that each were observed. Conversely, mates of 6 less successful pairs were simultaneously absent for periods of time ranging from 2.8–10.2 h (an average of 19% of the total time each pair was observed during incubation). Three of these pairs abandoned their clutches about one week before eggs were due to hatch. In one case, the female contributed most extensively to incubation; in the two other cases, the males spent a disproportionate amount of time incubating in both morning and evening periods (see Fig. 1B for two cases).

Parental Attendance with Chicks

During the period of chick care, the mean time spent in observation of the 27 pairs that hatched at least one egg was 23.5 ± 4.1 h. In 10 pairs of successful parents, the partner that incubated primarily in the morning (or evening) was also primarily present with the chicks during the same time period. In each case, the partitioning of parental care activity observed during incubation was maintained during chick care. Mates in the 7 other successful pairs also continued to exhibit the time committment patterns of parental care during chick care that they earlier exhibited during incubation. An example of the pattern of attendance by mates of successful pairs is shown for male partners during morning (AM) observation periods (Fig. 2A). Similar positive correlations (Spearman Rank Correlation Tests) were found for three other comparisons of the timing of incubation and solitary attendance with chicks. These were (1) male incubation PM vs. male attendance with chicks PM (r_s = +0.57, P < 0.05), (2) female incubation AM vs. female attendance with chicks AM (r_s = +0.74, P < 0.01), and (3) female incubation PM vs. female atten-

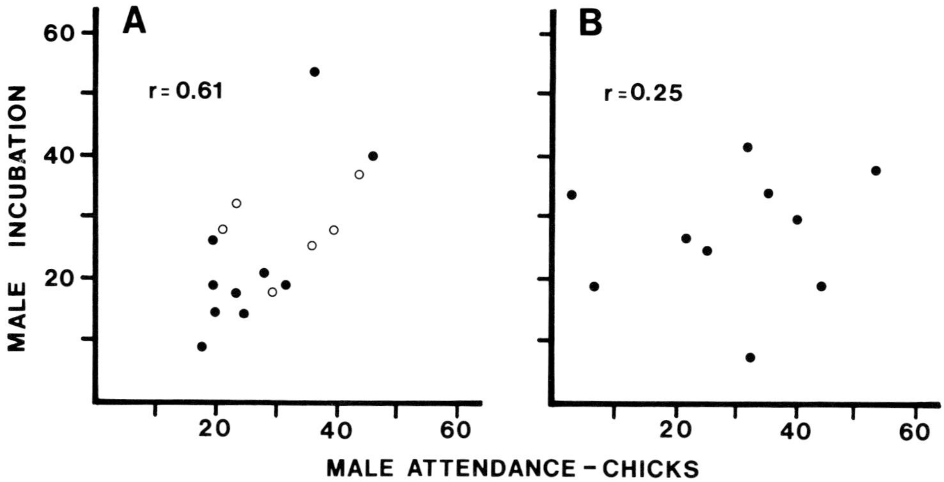

FIGURE 2. The relationship between time (mean minutes/hour) spent in incubation and time spent in solitary attendance with the chicks during the first 10 days of brood care. The example shown is for morning observation periods for male partners in Herring Gull pairs. A. Successful pairs (n = 16; incomplete incubation data available for one pair). B. Less successful pairs (n = 10; 4 pairs failed to hatch eggs). In A, closed circles are males of the pairs shown in Fig. 1; open circles are males from successful pairs where one partner exhibited somewhat more parental care activity than the other in both morning and evening time periods (see text for complete details).

dance with chicks PM ($r_s = +0.55$, $P < 0.05$). In each case, the correlations were based on 16 of the 17 pairs of successful parents; 1 pair had too few hours of observation during incubation for inclusion.

In contrast, these patterns were not observed among any of the less successful pairs for which data were available (n = 10). An example of the absence of correlation between incubation and attendance with chicks during morning observation periods is shown for males in less successful pairs (Fig. 2B). A similar lack of correlation was identified for two of the three further cases noted above (male-male PM, $r_s = +0.21$, $P > 0.05$; female-female AM $r_s = +0.22$, $P > 0.05$; female-female PM, $r_s = +0.57$, $P < 0.05$). In each case, correlations were based on the 10 pairs of less successful parents that hatched at least 1 egg.

DISCUSSION

The Herring Gull pairs observed in this study were segregated into two groups based on maximal differences in chick survival (see Burger 1986 for a similar procedure). Successful pairs realized a reproductive success rate (chicks fledged per pair) almost five times higher than that of less successful pairs. Less successful parents hatched fewer eggs, abandoned clutches, and lost chicks such that only 8 of 14 such pairs successfully raised a single chick. While these demographic measures of success were clearly different between the two groups, the pairs were similar in other respects. Females in all pairs laid 3-egg clutches during a colony-wide peak of egg laying early in the season each year. From other studies of larids, such individuals are known, on average, to represent an older, more experienced subset of birds in the colony (younger, less experienced larids produce smaller clutches later in the season). Some of the study animals were known to have bred in previous years; yet, despite other studies to the contrary (Coulson 1966, Mills 1973), there was no evidence among Herring Gulls at Port Colborne that "change" pairs were less successful than "same" pairs.

The patterns of parental attendance during incubation and chick care paralleled differences in the reproductive success of pairs in the two success categories. Successful partners were coordinated in their contributions to egg and chick care; less successful parents were not. Partners of 10 successful pairs were predictable in the time of day when care was given and were in synchrony with each other in the partitioning of time allotted to parental care activities. In the remaining 7 successful pairs, one mate spent somewhat more time incubating and attending chicks than the other. Rather than a "trade-off" pattern of parental care, partners exhibited general time equitability of care given to the clutch and brood. Equitability of time investment in incubation and chick care activities has elsewhere been suggested as characteristic of successful Herring Gull parents (Burger 1986). In either case, partners of successful pairs in this study coordinated their parental care behavior such that joint absence was infrequent and contributions to egg and chick care were synchronized.

Conversely, in less successful pairs, one partner exhibited disproportionate contributions to incubation and chick care, and was the sole partner in attendance during much of the day. There was no evidence of synchrony or coordination in the timing of care given to eggs or to chicks, and in several cases, both partners were frequently absent simultaneously from the clutch or brood.

Sexual selection theory predicts that the greater investing sex (usually females) should be more discriminating in the choice of a partner than the less investing sex (usually males; Orians 1969, Trivers 1972). Burley (1977) extended sexual selection theory to predict within-sex variability with regard to selectivity toward mates. She showed that while female feral pigeons (*Columba livia*) were relatively more selective of mates than males, high quality males were selective as well. Among colonially nesting seabirds, total effort given to parental care activities by males appears to be equal (Pierotti 1981, Western Gulls; Butler and Janes-Butler 1983, Great Black-backed Gulls, *L. marinus*), or exceed (Montevecchi and Porter 1980, Northern Gannets, *Morus bassanus*; Burger 1981, Black Skimmers, *Rynchops niger*) that of females. While differing on "parental investment" and "mating effort" terminology, these studies agree that overall parental care contributions by male and female partners are similar. Both parents care for eggs and chicks and are important to the reproductive fitness of each other. Accordingly, high quality in a mate is in the best interest of both partners, and selection pressure for choice of a high quality mate likely acts on both sexes.

While monogamous partners both contribute to the care of offspring, sexual differences in investment often take different forms and occur at different times during breeding. Predictions about which sex might be more discriminating in choice of a mate are complicated by difficulties in defining the currency of investment allocated at different phases of breeding (Knapton 1984). Among seabirds, "typical" male and female roles are usually well defined, but principal contributions occur at different times or in different ways. For example, female Great Black-backed Gulls invested more time than males in territorial attendance and incubation whereas, males engaged more in agonistic behavior during the post-hatch period (Butler and Janes-Butler 1983). Similarly, while the agonistic behavior of female Western Gulls toward intruders was less extensive than that of males, the timing of such acts by females was considered to represent an important component of territorial defense investment by the pair (Hand 1986).

Among seabirds, then, when extensive parental contributions are given by both parents, males and females are both likely to assess the quality of prospective partners. Differences in within-sex parental quality have been inferred by investigators from various indicators including clutch size (Coulson and Porter 1985), differential adult mortality rates during the breeding season (Burley 1985), and differential courtship and chick feeding rates (Wiggins and Morris 1986). Cues reflecting quality are likely also available. For example, courtship feeding rates by males appear to provide females with cues to the quality of potential mates (Nisbet 1973, Niebuhr 1981, Wiggins and Morris 1986). Foraging for food likely entails greater energetic costs than remaining at the colony (cf. Pugesek 1981), yet courtship feeding is the most direct way a male can contribute to the quality of eggs produced by a female (Smith 1980). Accordingly, males should be selective as to which female they will offer food. Cues about female quality might also be available for use by males. For example, the number of aggressive acts by a female toward intruders, and her willingness to remain on the territory, may serve as indices by which a male can assess her quality as a reliable mate.

The results reported here indicate that differential mate quality appears to be expressed in attendance patterns of each partner during incubation and brooding. Previous work at the Port Colborne Herring Gull colony showed that joint absence or poor attendance by one partner resulted in increased loss of eggs and young chicks (Morris and Black 1980, Schoen and Morris 1983). In the present study, the timing of incubation by partners of successful pairs was a reliable predictor of the timing of solitary attendance with chicks. Synchrony in the timing of incubation attentiveness may, therefore, indicate to both partners the future willingness or ability of a mate to participate in brooding and protecting chicks. Ability to achieve synchrony in the timing of crucial parental care behavior suggests high quality in both partners. In colonial nesting Herring Gulls where cannibalistic neighbors are a persistent threat, some degree of parental cooperation is necessary for the survival of eggs and chicks. Failure by one partner to exhibit either synchronous or equitable incubation attentiveness, or to offer more than token attention to the clutch, probably indicates low parental quality of that individual. Early abandonment of the clutch by the most heavily investing partner may be a probable tactic in such cases, especially for young birds for which many future breeding bouts are likely (see Pugesek 1981). It is noteworthy that clutch abandonment, by members of both sexes, was observed in three pairs about one week before eggs were due to hatch. In each case, one partner exhibited extensive periods of absenteeism during the 10–14 days after clutch completion while the "faithful" partner was incubating throughout much of the day (see Fig. 1B for 2 cases).

Coordinated parental care behavior, proposed here as characteristic of high quality parents, can be expressed in ways other than synchrony or equitability of time invested in clutch and brood care activities. Ability to achieve compatibility in foraging schedules (Niebuhr and MacFarland 1983), or in equitable resolution of conflicts over nest relief during incubation (Hand 1985), are also correlates that may be associated with high quality parents. However measured, the variability in parental quality identified among partners in my study was unlikely to be due solely to differences in age or in previous breeding experience. Indeed, study pairs were selected for similarity in clutch size and timing of clutch initiation in order to minimize these variables. In concert with studies to determine the overall role of the sexes in offspring care, detailed observations of particular pairs as reported here can provide insight into differential parental quality of the partners.

ACKNOWLEDGMENTS

Several people helped collect parental care data during this and related projects at the Port Colborne Herring Gull colony between 1979 and 1981. They were Michael Bidochka, John Chardine, Jina Chudzik, Walter Giesbrecht, George Melvin, Kevin Teather, and Ralph Schoen. Joanne Siderius tabulated much of the attendance data from 1980 and 1981. Jack Bonisteel and Pat Kerr (Lighthouse Keepers), Chuck Rutledge (Marine Operator) and Don and Jean Winn (residents of Port Colborne) provided their usual high level of logistic assistance. The manuscript benefited from critical reviews by J. Burger, R. G. Butler, J. L. Hand and an anonymous reviewer. Funding was generously provided by the Natural Sciences and Engineering Research Council of Canada (operating grant #A6298). My sincere thanks to all these individuals and organizations.

LITERATURE CITED

Brown, R. B. G. 1967. Breeding success and population growth in a colony of Herring Gulls and Lesser Black-backed Gulls *Larus argentatus* and *L. fuscus*. Ibis 109:502–515.

Burger, J. 1981. Sexual differences in parental activities of breeding Black Skimmers. Amer. Natur. 117:975–984.

Burger, J. 1986. Selection for equitability in some aspects of reproductive investment in Herring Gulls (*Larus argentatus*). Ornis scand. 18:17–23.

Burley, N. 1977. Parental investment, mate choice, and mate quality. Proc. Natl. Acad. Sci. USA 74: 3476–3479.

Burley, N. 1985. Leg-band color and mortality patterns in captive breeding populations of Zebra Finches. Auk 102:647–651.

Butler, R. G., and S. Janes-Butler. 1983. Sexual differences in the behavior of adult Great Black-backed Gulls (*Larus marinus*) during the pre- and post-hatch periods. Auk 100:63–75.

Chardine, J. W., and R. D. Morris. 1983. Herring Gull males eat their own eggs. Wilson Bull. 95:477–478.

Coulson, J. C. 1966. The influence of pair-bond and age on the breeding biology of the Kittiwake Gull *Rissa tridactyla*. J. Anim. Ecol. 35:269–279.

Coulson, J. C., and J. M. Porter. 1985. Reproductive success of the Kittiwake *Rissa tridactyla*: the roles of clutch size, chick growth rates and parental quality. Ibis 127:450–466.

Hand, J. L. 1985. Egalitarian resolution of social conflicts: a study of pair-bonded gulls in nest duty and feeding contexts. Z. Tierpsychol. 70:123–147.

Hand, J. L. 1986. Territory defense and associated vocalizations of Western Gulls. J. Field Ornithol. 57:1–15.

Knapton, R. W. 1984. Parental investment: the problem of currency. Can. J. Zool. 62:2673–2674.

Mills, J. A. 1973. The influence of age and pair-bond on the breeding biology of the Red-billed Gull, *Larus novaehollandiae scopulinus*. J. Anim. Ecol. 42: 147–162.

Montevecchi, W. A., and J. M. Porter. 1980. Parental investments by seabirds at the breeding area with emphasis on Northern Gannets *Morus bassanus*. Pp. 323–365 *in* J. Burger, B. L. Olla and H. E. Winn (eds.). Behavior of marine animals. Vol. 4. Plenum Press, New York.

Morris, R. D., and M. J. Bidochka. 1982. Mate guarding in Herring Gulls. Colonial Waterbirds 5: 124–130.

Morris, R. D., and J. E. Black. 1980. Radiotelemetry and Herring Gull foraging patterns. J. Field Ornithol. 51:110–118.

Morris, R. D., and J. W. Chardine. 1985. The effects of ice cover over the colony site in reproductive activities of Herring Gulls. Can. J. Zool. 63:607–611.

Morris, R. D., and G. T. Haymes. 1977. The breeding biology of two Lake Erie Herring Gull colonies. Can. J. Zool. 63:607–611.

Nisbet, I. C. T. 1973. Courtship feeding, egg-size and breeding success in Common Terns. Nature 241: 141–142.

Niebuhr, V. 1981. An investigation of courtship feeding in Herring Gulls *Larus argentatus*. Ibis 123: 218–223.

Niebuhr, V., and D. McFarland. 1983. Nest-relief behaviour in the Herring Gull. Anim. Behav. 31: 701–707.

Orians, G. H. 1969. On the evolution of mating systems in birds and mammals. Amer. Natur. 103: 589–603.

Parsons, J. 1971. Cannibalism in Herring Gulls. British Birds 64:528–537.

Pierotti, R. 1981. Male and female parental roles

in the Western Gull under different environmental conditions. Auk 98:532–549.

PUGESEK, B. H. 1981. Increased reproductive effort with age in the California Gull (*Larus californicus*). Science 212:822–823.

SCHOEN, R. B., AND R. D. MORRIS. 1983. Nest spacing, colony location, and breeding success in Herring Gulls. Wilson Bull. 96:483–488.

SMITH, S. M. 1980. Demand behavior: a new interpretation of courtship feeding. Condor 82:291–295.

SOUTHERN, L. K. 1981. Sex-related differences in territorial aggression by Ring-billed Gulls. Auk 98:179–181.

TRIVERS, R. L. 1972. Parental investment and sexual selection. Pp. 136–179 *in* B. Campbell (ed.). Sexual selection and the descent of man. Aldine, Chicago.

WIGGINS, D. A., AND R. D. MORRIS. 1986. Criteria for female choice of mates: courtship feeding and paternal care in the Common Tern. Amer. Natur. 128:126–129.

Studies in Avian Biology No. 10:75–79, 1987.

DO ADULT GULLS RECOGNIZE THEIR OWN YOUNG: AN EXPERIMENTAL TEST

JOSEPH G. GALUSHA AND RONALD L. CARTER[1]

ABSTRACT.—Adult gulls are reported to recognize their own chicks by the time they are 5 or 6 days old (Tinbergen 1953). We experimentally tested this hypothesis by allowing single chicks to live on their own territory for between 3 and 25 days before being switched to another territory. To prevent straying by the chicks, a chicken-wire fence, 30.5 cm high and enclosing 1 m^2, was placed around each nest at time of hatching. Twenty-two of 25 adult gulls accepted and offered food to chicks after experimental manipulation. Older chicks showed various degrees of "concern" during the first hour after being switched but they accepted food. Subsequently we rotated 10 chicks between 10 territories once each day for 30 days and then placed them on natural, control territories. These chicks begged for food, were fed and accepted in 8 of 10 trials. We conclude that Glaucous-winged Gulls do not recognize their own chicks individually.

In his now classic book, The Herring Gull's World (1953), Tinbergen reported that Herring Gull parents (*Larus argentatus*) recognized their young at five days of age and older. His primary evidence was that until this age, chicks could be exchanged between nests without adult rejection. After this time, chicks wandering naturally or placed experimentally in the vicinity of another nest were chased out of the territory or attacked and often killed.

The importance of individual recognition of chicks by parents is particularly germane to the sociobiological concept of inclusive fitness. To achieve maximal fitness for its own genes in the next generation, a parent gull not only must care for its own chicks efficiently, but must refrain from contributing to the survival of competing gene carriers at the expense of its own reproductive output (Miller and Emlen 1975). In closely-packed, territorial, gull colonies, theory indicates that natural selection should favor restriction of parental care to the "right" offspring, or mechanisms that maintain maximal parental attention to specific, healthy young (Hailman 1967).

Beer (1970a, 1979) performed several experiments showing that Laughing Gull chicks (*Larus atricilla*) could distinguish between the voices of their parents and other adults by as early as the first day of life. He further showed that by the end of the first week, they responded exclusively to the calls of their parents by approaching and calling but were silent, remained crouched or moved away during playback of calls of other adults. Miller and Emlen (1975) demonstrated that adult Ring-billed Gulls (*Larus delawarensis*) accepted any chick placed in or around their nest for the first five days after the chicks hatch, but rejected all intruders after day seven.

Holley (1981, 1984) reported naturally occurring adoptions of Herring Gull chicks between the ages of 6 and 35 days. From samples of three colonies studied in Great Britain, the rate of natural adoption was determined to be 13.4% (16 chicks of 119 nesting pairs). Several recorded adoptions were even between members of different but closely-related, sympatric species (Herring and Lesser Blackbacked gulls, *L. argentatus* and *L. fuscus*) breeding in the same colony. Of greater behavioral interest was the result of experimentally transferring 8 chicks to nests containing young of the same age. Four chicks were placed in territories from which they could easily escape; all were rejected by the resident adults. Another four were transferred to territories from which escape was not possible (due either to being on a cliff edge or within a small, fenced enclosure); all were accepted. Holley explains these results by noting that the chicks that could do so, tried to escape, a behavior resident chicks would not normally show. Thus they distinguished themselves as strangers. Those that could not escape showed only "appeasement behavior" (Holley 1984), with which the resident adults were familiar, and thus were not attacked.

The purpose of our study was (1) to experimentally confirm the age of parental recognition of individual chicks in Glaucous-winged Gulls (*Larus glaucescens*), reported to be about 5 days by Vermeer (1963); (2) to determine the effect of disrupting filial imprinting (by rotating chicks between different nests) upon the subsequent development of parent-young recognition; and (3) to determine if the behavior of chicks that had been rotated between several nests would allow them to be adopted more easily than natural, control chicks.

METHODS AND MATERIALS

Our experiments were conducted during the months of July 1983, 1984, 1985 on Protection Island, Jefferson County, Washington. This gull colony was composed of approximately 10,000 Glaucous-winged Gulls breeding on an area of about 25 ha.

At the beginning of each experiment, 10–25 nests

[1] Department of Biological Science, Walla Walla College, College Place, Washington 99324.

TABLE 1
RESULTS OF EXPERIMENT 1 IN WHICH CHICKS WERE RAISED IN AN ENCLOSURE BY ONE SET OF PARENTS FOR A VARIABLE NUMBER OF DAYS AND THEN SWITCHED TO AN ENCLOSURE ON A NEW TERRITORY

	Gull chicks		
Switch	Age on day of switch	Number switched	Number attacked
1	3	5	1
2	6	5	1
3	8	5	0
4	13	5	1
5	15	5	0
Total		25	3

TABLE 2
RESULTS OF EXPERIMENT 2 SHOWING NUMBER OF ATTACKS BY ADULTS ON 6 CHICKS WHICH WERE RAISED BY THE SAME PARENTS (STATIONARY CHICKS AND ADULTS) AND WERE EXCHANGED WITH 6 CHICKS WHICH WERE ROTATED ONE NEST PER DAY (ROTATED CHICKS AND ADULTS) AT SPECIFIED AGES

	Attacks per number switched by	
Age (days)	Stationary adults on rotated chicks	Rotated adults on stationary chicks
5	0/6	0/6
10	0/6	0/6
15	0/6	0/6
21	0/6	0/6
Total	0/24	0/24

containing three eggs were selected within 50 m of our blind. A few days before the first eggs were to hatch, each nest was marked with a numbered stake and surrounded with a chicken-wire fence (2.5 cm mesh) which was 30.5 cm high and enclosed a 1 m^2 area (hereafter called an "enclosure"). Also placed within 25 cm of the nest was a piece of PVC tubing (25 cm long × 10 cm in diameter and enclosed at one end) which served as additional shelter. The evening before each experiment, all eggs were removed and replaced with one pipping egg or freshly hatched chick at each nest.

EXPERIMENT 1 (1983)

In the morning (between 0800–1100 h) on day 3 and at approximately 3-day intervals until day 15, five chicks that had been raised singly by pairs of resident adults were randomly exchanged with one another (25 chicks total). Each chick was put into a new enclosure only once during the experiment. Observations and video records were made continuously for 6 h following each transfer. The occurrence of attacks, feedings, parent calls, and the presence or absence of other parent-young behavior was noted.

EXPERIMENT 2 (1984)

Twelve nests were used. Six, designated experimental, had their chicks rotated among them each day for 30 days. The other nests, designated controls, had their chicks carried around the colony each day for a duration similar to that required to move the experimental chicks (approximately 30 seconds) and then returned to their own enclosures. For the next two hours, note was made of whether the returning adult(s) attacked, fed, and/or called to the chicks or not.

On days 5, 10, 15, and 21 the six experimental chicks were randomly exchanged with the six control chicks and all chicks were left in new enclosures for 4 h. Observations mentioned above were then made. The chicks were then returned to their original nests and enclosures.

EXPERIMENT 3 (1985)

Chicks from ten nests were rotated one nest position each day for 30 days as in Experiment 2. On day 30, each of these chicks was carried in its shelter to a randomly selected natural nest located at least 20 m away with 2–3 resident chicks of approximately the same age (25–30 days). For the next 2 h we observed the interaction of each chick with its new family. Five of the experimental chicks were then returned to their enclosures for the night. The next day these chicks were removed and five randomly chosen, naturally reared chicks selected from the colony were placed in these enclosures and observed for 2 h.

RESULTS

EXPERIMENT 1

During 25 opportunities for acceptance or rejection by territorial adults of chicks transferred at 3–15 days, only 3 times were chicks attacked by adult gulls (Table 1). Interestingly, each attack occurred on a different day (3, 6, and 13). All other chick transfers resulted in parental acceptance (feeding, calling, adult care, etc.).

EXPERIMENT 2

In 48 trials using chicks, aged 5–21 days, which had been rotated daily or had remained with the same adults before being exchanged, no attacks were recorded (Table 2).

EXPERIMENT 3

Eight of ten chicks rotated daily for 30 days before being transferred to a natural nest in the colony were accepted into new families. In the two cases in which chicks were rejected, the transferred chick was first harassed by resident chicks and then attacked by the adults.

Three of 5 control chicks (25–30 days of age) placed in enclosures on territories of adults that had been subjected to chick rotation every day for the prior 30 days were also accepted.

OBSERVATIONAL EVIDENCE

Qualitative evidence from our field notes suggested that adult gulls did not behave differently toward young placed in their enclosure compared to their own chicks. For instance, during the first week, adult gulls jumped into the enclosures to brood and feed experimental and control chicks alike. These adults also "mewed" (Tinbergen 1959) around the outside periphery of the fences for minutes at a time, occasionally regurgitating food in front of the chick. Often the chick attempted to eat the food but, because of the fence, was unable to do so. Other times the adult held the food in its bill and the chick ate. Occasionally, especially when older, the chick ran to the other side of the enclosure at these times and attempted to get out. The resident gulls then alarm-called (*kek* and *keow*; Tinbergen 1959), re-swallowed the deposited food, "mewed" around the fence and offered food again. Inevitably, however, even these chicks would eat. We saw this sequence occur with both experimental and control chicks.

Behavior of control and experimental chicks was consistently very similar. After being switched to new enclosures, chicks hid in the tube shelters until the colony disturbance subsided. Then they backed out. If the chicks were less than a week old or had been rotated daily between other enclosures, they stood with their necks extended in an erect posture and looked around. When older, a few switched chicks tried to escape from what was a new area to them for extended periods of time. The chicks might then explore their new enclosure before begging for food, which when offered by their new parents, was eaten through the mesh holes in the fence. Chicks which had been raised by the same parents for extended periods of time (10 days or more) occasionally cowered in the corners of the new enclosures for a considerable time after being switched. It was our impression that they behaved in a less "confident" way than did the rotated chicks or chicks which were returned to their own territory.

DISCUSSION

There was no consistent evidence that adult Glaucous-winged Gulls differentiated between strange chicks they had not seen before and individual chicks that they had raised for periods of time up to 30 days. In fact, our data strongly suggests that adult gulls do not recognize chicks individually. This finding is similar to Holley's (1984) conclusion that adult Herring Gulls are unable to recognize their own young below the age of 35 days. Beer (1970b) has also reported several instances where adult Laughing Gulls attacked their own young or adopted unrelated chicks.

If adult Glaucous-winged Gulls do not distinguish between their own and other chicks on the basis of visual, morphological cues, why is it that they regularly chase, attack and kill chicks who enter their territory? We presently believe that the best explanation supported by our data is that adult gulls of this species, and probably others, identify their young, biological or foster, primarily by the way they behave. That is to say, as long as a chick's behavior is appropriate for an "offspring," it will be treated as such by the resident adults. Miller and Emlen (1975) referred to this possibility as "comportment discrimination." Beer (1979) hinted at a similar, possible interpretation when he offered that Tinbergen's earlier observations of apparent parental recognition of chicks might, in fact, be due to recognition of parents by chicks, with consequent changes in chick behavior being used by adults to discriminate between their own and unrelated chicks.

Chick Recognition by Adults

The strongest experimental evidence to date suggesting that adult gulls might distinguish their young individually by visual cues is the work of Miller and Emlen (1975) on Ring-bills (*L. delawarensis*). Of 12 chicks whose head plumage and facial features were altered with black ink, over half were attacked or at least not accepted. It may be important to know the extent and similarity of the alterations before drawing further conclusions, for the location or configuration of the markings may have obliterated the "gull chickness" of these chicks. That they were initially rejected but eventually accepted (within a couple of hours) and not killed suggests that they were treated differently than natural chick intruders.

Other chick exchanges by Miller and Emlen demonstrated differential rejection of chicks between the ages of 5 days (1 rejected/14 presented), 7 days (7 rejected/14 presented) and 9 days (10 rejected/14 presented). One cannot rule out the possibility that chick behavior changed during this time period and caused the consequent change in adult behavior.

Beer (1979) has shown that adult Laughing Gulls respond to the recorded voices of chicks by orienting toward, calling and approaching the source of the sound. These same adults did not discriminate, however, between the calls of their own and foreign chicks. Beer concluded that it is very unlikely that parent gulls of this species can recognize their chicks individually by voice.

Adult Recognition by Chicks

Currently there is more evidence to support the idea of recognition of parent or territory by chicks than individual recognition of chicks by adults. For instance, Beer (1970a, 1979) and Miller and Emlen (1975) have shown that chicks of Laughing and Ring-billed gulls can differentiate the calls of their parents from other adults by the end of the first week post-hatching. As mentioned above, however, parents did not distinguish the calls of their own chicks.

As yet we have not been able to do the quantitative analysis of our chick data necessary to compare differences between the behavior of resident and strange chicks on a natural territory. We believe, however, that our data will support previous work showing that chicks act differently when off, compared to on, home territories. When at home, chicks spend most of their undisturbed time standing with their necks contracted against their bodies (like at rest) or erect and alert. Chicks that are disturbed or off their natural home territory spend much of their time crouched down in cover or running from place to place searching for a place to hide. During exchanges of older chicks in our experiments, we observed these "disturbed" behaviors but not nearly as often as by naturally displaced chicks.

Noseworthy and Lien (1976) have shown that the behavior of chicks not on their territory changes with age. Specifically, the numbers of chicks that return to their nest site after being experimentally moved a standard distance increases during the first week post-hatching until it reaches a maximum at 6 or 7 days. Older chicks do not return as often. If return rate correlates with time of maximal attachment, as they contend it does, then another factor influencing normal chick behavior might be recognition that they were or were not on their natural territory.

To reduce the effect of chick recognition of new territory after being switched, we confined the control and experimental chicks to 1 m^2 of surface area around the nest via fenced enclosures. We also increased topographical homogeneity in the enclosures by removing tall weeds and vegetation. Though there is still a possibility that the chicks recognized that they were not at home, our efforts may have reduced the dissonant cues perceived by both experimental and control chicks and contributed to their acting quite similarly, and naturally, even when switched to new territories and enclosures.

Gull chicks are also known to differentiate their siblings (at least nestmates) from other chicks at a very early age (Noseworthy and Lien 1976). To reduce this potentially confounding variable, we used single chicks raised alone in our experiments. Though we are unaware of any reasons why this might have been inappropriate, it may have affected the development or extent of parent-young attachment.

Another possible explanation of our findings must be considered. It is related to the argument of Graves and Whiten (1980) that adoptions may occur in gull colonies due to the apparent inhibition of adults to attack chicks in close proximity to the nest. As all of our fences enclosed areas around natural nest sites, the consistent acceptance of chicks under our experimental conditions might be thus explained. We have begun work with enclosures placed at different locations in the territory to further evaluate this hypothesis. Also, the mere presence of the fences may have reduced the willingness of adults to attack chicks on their territory. Future experiments using chicks tethered at various locations on the territory will contribute to the answer of this question.

Finally, we do not yet have experimental data from members of biologically related and unrelated families bearing on the question of kin vs. non-kin recognition by gulls. This is an important theoretical extension of our work and is a direction we shall pursue in the future.

SUMMARY

Resident gulls fed and cared for single experimental chicks placed in fenced enclosures surrounding their nests from time of hatching. When chicks were exchanged among nests at ages between 3 and 15 days, 22 of 25 were accepted and cared for within 2 h. Chicks rotated daily between several nests and then placed in control nests, or in enclosures where chicks had been rotated daily, were not attacked by resident adults. Eight of 10 chicks rotated daily between different nests for 30 days were accepted into new families. Even 3 of 5 natural chicks which had been with their biological parents for 25–30 days were accepted by adults which had had their chicks rotated. We conclude that adult gulls do not recognize their chicks individually but treat them as their "own" or "strangers" based upon how they behave. Factors such as recognition of home territory, siblings, and parental voice may influence how chicks act, which in turn determines how they are treated by resident adults.

ACKNOWLEDGMENTS

Special appreciation is due Mr. Phillip Vorvick, President of the Protection Island Beach Club, and Mr. John Lorenz, Caretaker, for permission to conduct this research and for logistic support on Protection Island, Jefferson County, Washington. We are especially grateful for the provision of comfortable living accommodations so generously provided by Mr. and Mrs. Marion R. Burrell.

This research was supported by a Walla Walla College Faculty Grant to the first author. Generous financial support was also provided by the Walla Walla College Marine Station, Anacortes, Washington.

We wish to thank the editors and two anonymous reviewers for their careful reading and comments which greatly improved earlier versions of this paper.

LITERATURE CITED

BEER, C. G. 1970a. On the response of Laughing Gull chicks (*Larus atricilla*) to the calls of adults I. Recognition of the voices of parents. Anim. Behav. 18: 652–660.

BEER, C. G. 1970b. Individual recognition of voice in the social behavior of birds. Pp. 27–74 *in* D. S. Lehrman, R. A. Hinde, and E. Shaw (eds.). Advances in the study of behavior, Vol. 3. Academic Press. New York.

BEER, C. G. 1979. Vocal communication between Laughing Gull parents and chicks. Behaviour 70: 118–146.

GRAVES, J. A., AND A. WHITEN. 1980. Adoption of strange chicks by Herring Gulls (*Larus argentatus*). Z. Tierpsychol. 54:267–278.

HOLLEY, A. J. F. 1981. Naturally arising adoption in the Herring Gull. Anim. Behav. 29:302–303.

HOLLEY, A. J. F. 1984. Adoption, parent-chick recognition and maladaption in the Herring Gull (*Larus argentatus*). Z. Tierpsychol. 64:9–14.

HAILMAN, J. P. 1967. The ontogeny of an instinct. Behavior Supplement 15.

MILLER, D. E., AND J. T. EMLEN. 1975. Individual chick recognition and family integrity in the Ring-billed Gull. Behaviour 52:122–144.

NOSEWORTHY, C. M., AND J. LIEN. 1976. Ontogeny of nesting habitat recognition and preference in neonatal Herring Gull chicks (*Larus argentatus*) Pontappidan. Anim. Behav. 24:637–651.

TINBERGEN, N. 1953. The Herring Gull's World. Collins, London.

TINBERGEN, N. 1959. Comparative study of the behavior of gulls (Laridae): a progress report. Behaviour 15:1–70.

VERMEER, K. 1963. The breeding biology of the Glaucous-winged Gull (*Larus glaucescens*) on Mandarte Island, B. C. Occas. Pap. Brit. Colum. Prov. Mus. 13:1–104.

Studies in Avian Biology No. 10:80, 1987.

A SIMULATION MODEL OF FLOCK FORMATION IN RING-BILLED GULLS

ROGER M. EVANS[1]

ABSTRACT.—Previous studies suggest that flocks of Ring-billed Gulls (*Larus delawarensis*) arise from social facilitation of flight superimposed upon otherwise random departure times. A simulation model, based on these assumptions, is developed employing three variables (1) probability of being facilitated, (2) amount of time facilitated, and (3) time span over which facilitation can occur. Simulations using a range of values for these variables were compared with results of 12 actual data sets. A wide range of *P* of following provided a close fit to the actual data, but the amount of time facilitated rarely exceeded 80% of the time separating successive birds, as expected for gulls that "straggle" out in loose flocks. The best fit to the data was when facilitation acted over the biologically reasonable time of about 1 min. I conclude that the three variables modelled provide a realistic first approximation in this species.

[1] Department of Zoology, University of Manitoba, Winnipeg, Manitoba R3T 2N2, Canada.

Studies in Avian Biology No. 10:81, 1987.

SPATIAL AND TEMPORAL ASPECTS OF FRANKLIN'S GULL FLOCKS

J. G. KOPACHENA[1]

ABSTRACT.—Franklin's Gulls (*Larus pipixcan*) arriving and departing from a breeding colony were clumped in space and time. Daily changes in directional use and the number of directions used declined as the breeding season advanced. Temporal clumping was greatest in the evening. Arrivals were often more clumped and had shorter flock durations than departures. Most flights showed more than one pattern of flocking. Interbird distances were greatest for morning departures and afternoon arrivals. An increased temporal clumping of birds away from the colony, linear flight paths of departing individuals, and high call frequencies of birds in flight, suggests that social facilitation superimposed on random departure times is one causal explanation of flock formation. Variation in spatial and temporal clumping patterns indicates that the extent to which arrivals and departures are socially facilitated is associated with the time of day and flock destination.

[1] Department of Zoology, University of Manitoba, Winnipeg, Manitoba R3T 2N2, Canada.

Studies in Avian Biology No. 10:82, 1987.

NEIGHBOR INTERACTIONS AND COOPERATION AMONG BREEDING HERRING GULLS: AN ALTERNATIVE INTERPRETATION OF GULL TERRITORIALITY

GARY W. SHUGART[1] AND MARY A. FITCH[2]

ABSTRACT.—Theoretical studies predict that cooperation may evolve when individuals interact repeatedly and when costs and benefits to pairs of interactants are equivalent. These conditions apply to neighbors in territorial birds. One might expect neighbors in Herring Gulls (*Larus argentatus*) to be uncooperative because of the relatively high rate of chick killing by neighbors that has been reported in the literature. This mortality results from movement of chicks from parents' territories. Chick movement is a form of defection in a stable aggressive system based on site fixity and site-specific dominance. In previous studies of aggression in gulls, chick mortality probably has been exaggerated because of defections induced by human disturbance. A test of the above prediction regarding cooperation among gulls requires an examination of neighbor interactions in the absence of disturbance. In four years of study, we found that (1) mortality due to neighbors is rare in the absence of disturbance and (2) aggressive interactions between neighbors provides evidence for our hypothesis that cooperation between neighbors is an evolutionary force in coloniality.

[1] Department of Psychology NI-25, University of Washington, Seattle, Washington 98195.

[2] 5230 16-NE, Seattle, Washington 98105.

Studies in Avian Biology No. 10:83–90, 1987.

FORAGING EFFICIENCY IN GULLS: A CONGENERIC COMPARISON OF AGE DIFFERENCES IN EFFICIENCY AND AGE OF MATURITY

JOANNA BURGER[1]

ABSTRACT.—I studied feeding efficiency in 15 species of gulls in North America, Africa, Australia, and Europe in different feeding situations and habitats to test the hypotheses that (1) foraging efficiency increases with age, and that (2) the disparity between young and adults is greater in species with the most delayed maturity. Efficiency was measured by timing the interfood interval (time between successful food captures) for actively feeding individuals. Regression analysis indicated that the variance in the interval was explained by species, age, food type, method, and habitat. Adult interfood intervals were shorter than those of young for all species in all but 5 feeding situations. In general, the disparity between adult and young was greater for the large gulls, supporting the hypothesized relationship between deferred maturity and foraging efficiency of gulls.

All birds are faced with the task of finding food resources to provide adequate food for themselves throughout the year, and for developing young during the breeding season. Foraging theory suggests that birds should concentrate their foraging efforts where there is abundant, easily obtained, high quality food (Schoener 1971, Krebs et al. 1981). For most seabirds, food is patchily distributed, distant, and ephemeral (Ashmole 1963, Diamond 1984). Seabirds have some of the largest foraging ranges of any vertebrate (Dunnet and Ollason 1982, Duffy 1983, Ainley et al. 1984). Further, many seabirds forage primarily by plunge-diving for food, a difficult task for young to perfect (Ashmole 1971). The difficulties of learning where to forage and how to plunge-dive are reflected in age-related differences in foraging: young are significantly less efficient in many aspects of foraging behavior (Orians 1969, Searcy 1978). Age-related differences in feeding ability have been noted for seabirds such as gulls *Larus* (Verbeek 1977, Ingolfsson and Estrella 1978, Searcy 1978, Ulfstrand 1979, Burger and Gochfeld 1981, 1983; MacLean 1986), terns *Sterna* (Dunn 1972, Buckley and Buckley 1974), pelicans *Pelecanus occidentalis* (Orians 1969), frigatebirds *Fregata magnificens* (Gochfeld and Burger 1981), and cormorants *Phalacrocorax olivaceus* (Morrison et al. 1978). In all cases adults had lower interfood intervals, higher foraging success, spent less time foraging for, obtaining, and handling food items, foraged for less of the day than young, or exhibited a combination of these factors. These age-related differences have been cited as a cause for delayed maturity in seabirds, many species of which do not breed until their fourth year or later (Lack 1967, Ryder 1980).

Although many groups of seabirds have diets mainly limited to fish or other pelagic organisms, the *Larus* gulls have diversified in foraging method and habitat. Gulls forage in a variety of habitats from wet fields to the open ocean, and employ a variety of feeding methods appropriate to habitat and prey type (Hunt and Hunt 1973, Mudge and Ferns 1982). Further, they exhibit species-related differences in maturation from species that start to breed in their second year to those that breed in their fourth or even fifth year (Dwight 1925). Thus, gulls provide an excellent opportunity to study species, habitat, method, and age-related differences in foraging behavior among congenerics, and I do so in this paper. I use the interfood interval as an index of foraging efficiency, defining it for an actively foraging bird as the time between obtaining one item and obtaining a second food item.

Based on optimal foraging theory and the variability of gulls in size, age of sexual maturity, feeding methods, and prey items, I made the following predictions.

1. For all foraging situations and prey items adults should have shorter interfood intervals than young (see references above).
2. The ratio of young to adult interfood interval should increase with increasing difficulty of the task as measured by adult interfood interval (see Burger and Gochfeld 1983).
3. The ratio of young to adult interfood interval should relate directly to size and age of sexual maturity. That is, gulls that do not breed until they are four or five years old should exhibit a greater age-related difference in interfood interval than those breeding in their second year.

I tested these hypotheses by examining the interfood intervals of 15 species of gulls foraging in a variety of habitats in North America, South America, Europe and Australia. Table 1 lists the species examined with their scientific names.

METHODS

From 1977 to 1985, I collected data on interfood intervals on gulls wherever and whenever there was a

[1] Department of Biological Sciences and Bureau of Biological Research, Rutgers University, Piscataway, New Jersey 08855.

TABLE 1
DATA SETS, BODY LENGTH AND AGE OF FIRST BREEDING FOR *LARUS* GULLS STUDIED (1977–1985)

Gull species	Scientific name[a]	Number of data sets[b]	Mean length (cm)[c]	Age at first breeding[c,d] (year)
Heermann's	*Larus heermanni*	3	46–53	3
Ring-billed	*L. delawarensis*	12	45–53	3
Mew[d]	*L. canus brachyrhynchus*	11	40–46	3
Common[d]	*L. c. canus*	3	40–46	3
Herring	*L. argentatus*	21	56–66	4
California	*L. californicus*	1	51–58	3–4
Western	*L. occidentalis*	11	61–68	4
Kelp	*L. dominicanus*	1	58	4
Great Black-backed	*L. marinus*	2	71–79	4–5
Glaucous-winged	*L. glaucescens*	8	61–68	4
Laughing	*L. atricilla*	6	38–43	3
Grey-headed	*L. cirrocephalus*	2	41–43	3
Silver	*L. novaehollandiae*	1	38–43	2
Black-headed	*L. ridibundus*	8	38–43	2
Bonaparte's	*L. philadelphia*	9	33–36	2
Black-legged Kittiwake	*Rissa tridactyla*	7	39–46	2

[a] Order follows Morony et al. 1975.
[b] A data set usually consists of at least 20 interfood intervals each for adults and young of the year.
[c] From Harrison 1983, 2 years = 27–28 months old, 3 = 39–40 months old, 4 = 51–52 months old.
[d] Common and Mew are conspecific, but have non-contiguous ranges.

group including adults and young. The study was primarily coastal and did not involve observations from large, offshore vessels. Data were collected on all actively foraging gulls encountered whenever the flock met the following criteria: there were at least 20 individuals each of adults and young foraging at close range so food items could be identified, and success determined. Data were collected from 1977 through 1985 in the following locations for the following species: Heermann's (California, USA), Ring-billed (California, Texas, and New Jersey, USA; Alberta, Canada), Mew (=Common, California and Alaska, USA), Herring (Slimbridge, England; Ekkeroy, Norway; Leningrad, Tbilisi, and Baku, USSR), California and Western (California, USA), Kelp (Durban, South Africa), Great Black-backed (New Jersey, USA), Glaucous-winged (California and Alaska, USA), Laughing (New Jersey, Texas, USA), Grey-headed (Naivasha, Kenya), Silver (Brisbane, Australia), Black-headed (Guayaquil, Ecuador; Naivasha, Kenya; Moscow and Leningrad, USSR; Slimbridge, England), Bonaparte's (California, Alaska and New Jersey, USA; Alberta, Canada), Black-legged Kittiwake (Ekkeroy, Norway; Alaska, USA).

Data collected at each field site included: place, date, time, habitat, species, feeding method, and food. Habitat types included ocean, bay, mudflat, pond, lake, river and field. Food types included natural foods such as midges (Chironomidae), other insects, fish, worms, crabs, and other invertebrates, and man-made food such as offal (e.g., fish parts from canneries) and garbage. Except for small invertebrates picked from the water's surface, the food type could be easily identified. Feeding methods (modified after Ashmole and Ashmole 1967, Ashmole 1971) included surface-diving, plunge-diving, picking up items from the ground, piracy, aerial dip, surface dip and jump dive. Dipping could thus occur by flying or hovering above the water, or by swimming and dipping with the head down. If a bird flew up 20 to 50 cm and dove into the water it was called a jump dive. "Adult" was defined as adult-plumaged birds with all white tails, and "young" was defined as birds that were less than one year old (i.e., had hatched the last breeding season). The few birds in intermediate age classes were excluded from this paper.

In almost all cases a data set includes data on twenty adults and twenty young of one species. If feeding ceased during the observation period the data set might have fewer observations. On a randomly chosen individual, I recorded the interfood interval to be used as a measure of foraging ability. The interfood interval is the time between first obtaining a food item and successfully obtaining a second food item. I then switched to another individual, alternating adult and young when possible.

During the study I recorded interfood intervals for 4502 birds in 106 data sets (Table 1). Most species were examined from more than one location (see above) and habitat. Table 1 also lists the mean length of gulls and the usual age of first breeding.

I used stepwise multiple regression model procedures to determine the factors that should be entered in the model contributing to explaining the variance in interfood intervals. Ordinal variables could be analyzed without transformation. For each non-ordinal variable, I created a new variable which compared the dependent measures at each variable value against all other variable values, examining each value in turn. Thus for age of maturity, those that mature at 2 years of age were compared to those maturing at all other ages; then those that mature at 3 years were compared to all others, and so on. If any one of the values was significant it was entered in the model (see Burger et al. 1984). The stepwise procedure first selects the variable that contributes the most to the coefficient of determination (R^2), and then selects the second variable

TABLE 2
REGRESSION MODELS FOR INDIVIDUAL SPECIES EXPLAINING VARIATION IN INTERFOOD INTERVALS FOR A VARIETY OF FEEDING SITUATIONS

Species	Model				Factors entering models			
	F	R^2	df	*P*	Method	Food	Habitat	Age
All Species[a]	61.5	0.70	126,4376	0.0001	53.9 (0.0001)	291.4 (0.0001)	15.7 (0.0001)	37.9 (0.0001)
Heermann's	23.9	0.51	2,46	0.0001	29.3 (0.0001)	—	—	18.6 (0.0001)
Ring-billed	29.3	0.44	13,419	0.0001	24.0 (0.0001)	23.0 (0.0001)	14.6 (0.0001)	29.5 (0.0001)
Mew	32.9	0.50	15,500	0.0001	13.5 (0.0001)	75.9 (0.0001)	8.81 (0.003)	28.4 (0.0001)
Common	26.1	0.59	4,73	0.0001	33.3 (0.0001)	5.73 (0.01)	—	32.3 (0.0001)
Herring	25.9	0.51	24,599	0.0001	26.6 (0.0001)	55.8 (0.0001)	7.62 (0.0001)	39.11 (0.0001)
California	2.41	0.09	2,51	NS				
Western	25.8	0.45	14,439	0.0001	16.8 (0.0001)	45.6 (0.0001)	39.8 (0.0001)	—
Kelp	9.84	0.40	2,14	0.0001	—	—	—	9.84 (0.001)
Great Black-backed	0.18	0.01	1,14	NS				
Glaucous-winged	13.6	0.40	9,184	0.0001	20.2 (0.0001)	—	—	8.13 (0.0001)
Laughing	13.7	0.32	5,144	0.0001	—	32.9 (0.0001)	—	—
Grey-headed	16.8	0.48	3,59	0.0004	—	—	—	16.8 (0.0004)
Silver	37.83	0.63	2,35	0.0001	31.7 (0.0001)	—	—	44.2 (0.0001)
Black-headed	48.2	0.62	8,23	0.0001	36.5 (0.0001)	120.8 (0.0001)	—	16.5 (0.0001)
Bonaparte's	15.7	0.33	9,291	0.0001	25.3 (0.0001)	15.9 (0.0001)	—	10.7 (0.001)
Black-legged Kittiwake	43.1	0.57	8,257	0.0001	22.0 (0.0001)	—	—	3.02 (0.05)

[a] For this model, species (F = 86.2, $P < 0.0001$), foraging plane (F = 2.26, $P < 0.05$), and species × age (F = 70.7, $P < 0.0001$) also entered the model.

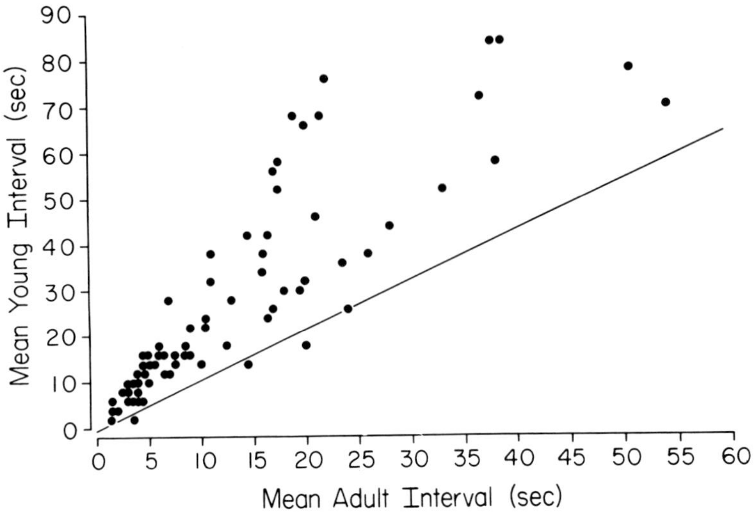

FIGURE 1. Relationship of mean interfood interval of young to mean interfood interval of adult gulls. Shown also is the line of equivalency where the mean interval for adults = the mean interval for young.

that gives the greatest increase in R^2. This procedure continues until all variables not included are not statistically significant ($P > 0.05$). If variables are highly correlated, only the one giving the highest R^2 is added. The model selection process determines the "best" model, gives the R^2 values and levels of significance for the model as well as for each contributing factor. I used Kruskal-Wallis χ^2 tests to determine significant differences in interfood interval as a function of age for each data set. I used analyses of variance to determine differences among classes of variables. All tests were performed on log-transformed data.

RESULTS

Factors Influencing Foraging Efficiency

I used multiple regression techniques to determine the factors influencing the variability in interfood interval (see methods). Over 60% of the variance in interfood interval for the 4502 individuals examined in this study was explained by (in order of contribution): food type, species, species × age, foraging method, age, habitat, and foraging plane (Table 2). Thus, food type was more important than species in explaining the variance in the interfood interval (see below).

I then examined the relative contribution of the variables to the variance in interfood interval for each species (Table 2). For all species except California and Great Black-backed gulls, the models explained a significant amount of the variation in interfood interval. This may be due to having only one and two data sets for the latter two species, respectively (thus there could be only one or two differences in the class variables such as method, food type and habitat).

The models indicate several things: (1) Between 30 and 60% of the variability in interfood interval was explained by the models. (2) Age was a significant factor for 12 of the 13 significant species models, but did not enter for Laughing Gull. (3) Method, food, and habitat entered fewer

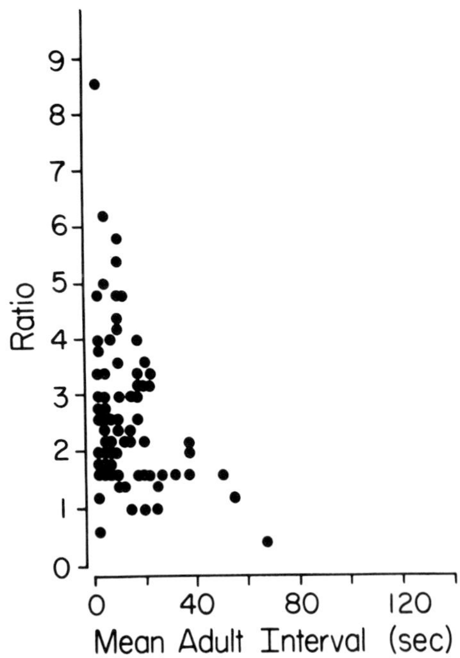

FIGURE 2. Relationship of rate of mean young and mean adult interfood interval to mean adult gull interfood interval for each of 106 data sets.

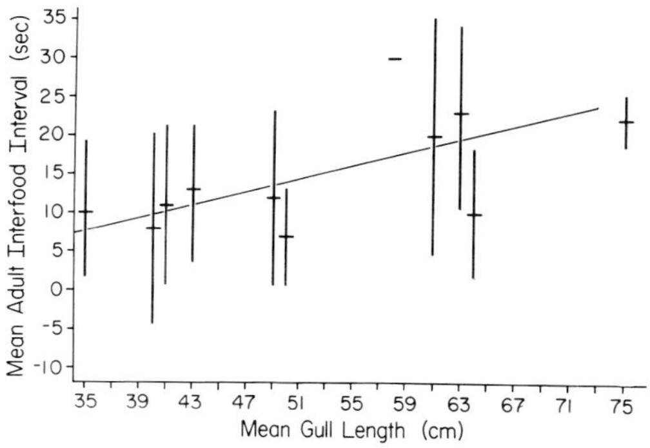

FIGURE 3. Relationship of mean interfood interval to mean body length for gulls.

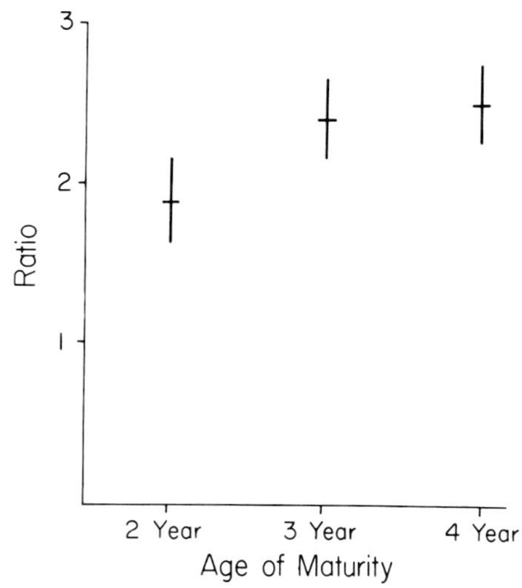

FIGURE 4. Mean foraging ratio (young interfood interval/adult interfood interval) as a function of body length in gulls.

of the significant models. Thus, for most species, age was one of the most important contributors to explaining variation in the interfood interval.

Effect of Age on Interfood Interval

Adults had significantly shorter interfood intervals in 96 of these data sets (91%). These results generally support prediction one: adults have lower interfood intervals than young. Exceptions were Great Black-backed Gulls feeding on garbage, Herring Gulls picking up worms, Laughing Gulls picking up insects from a wet field, Black-headed Gulls aerial dipping for invertebrates on a bay, and Bonaparte's Gulls picking up insects from a mudflat, plunging for fish and dunking for invertebrates on a lake. The exceptions are of interest since all of them fed on natural foods except the Herring Gull and were among the gulls that mature earliest.

The interfood interval for young was positively related to the adult interval (Kendall tau = 0.71, df = 210, $P < 0.0001$, Fig. 1). The ratio of young to adult foraging interval, however, was inversely related to the length of the foraging interval (Kendall tau = −0.18, df = 212, $P < 0.007$, Fig. 2). Thus, as the interval gets longer, the difference between adult and young decreases rather than increases, and prediction 2 is rejected.

Interfood Interval, Body Size and Age of Maturity

In gulls, age of maturity is related to body length: larger gulls mature and breed at a later age than smaller gulls (see Table 1, tau = 0.76, n = 16, Z = 4.09, $P < 0.001$). For the gulls studied, mean interfood interval increased as body size increased (F = 5.79, $r^2 = 0.39$, df = 1,9, $P < 0.03$, Fig. 3). The ratio of young to adult interfood interval increased significantly with increasing gull length (F = 4.23, $r^2 = 0.42$, $P < 0.0002$, Fig. 4). That is, as gull size increased the disparity between the mean interfood interval of adults and young increased, confirming prediction 3.

Although the gulls range in length from 33 to 79 cm, they can be grouped into three "maturity age" classes. The small gulls first breed when they are two years old, the intermediate-sized gulls breed when they are three, and the large gulls breed when they are four or five years old. For the 106 data sets examined the disparity between adult and young interfood interval differed significantly among the three "maturity age" classes (F = 4.12, df = 2, $P < 0.01$, Fig. 5). The difference between species with a two year vs. three year cycle was significant ($P < 0.05$) as were the two vs. four-year species ($P < 0.01$). There was no significant difference between species with three and four year cycles. Thus the interfood interval of young gulls that will breed in their second year was closer to the interfood intervals of adults than was the young to adult ratio for the larger gulls that breed in their fourth or fifth year, confirming prediction 3.

DISCUSSION

Methodology Problems

In this study, I examined only foraging situations where both adults and young foraged together. I selected these situations because all birds could theoretically feed on the same species and size of prey items, using the same methods. Although adults frequently fed in other foraging situations without young, it would bias the data to compare adult rates in these situations with young foraging in others. By selecting foraging

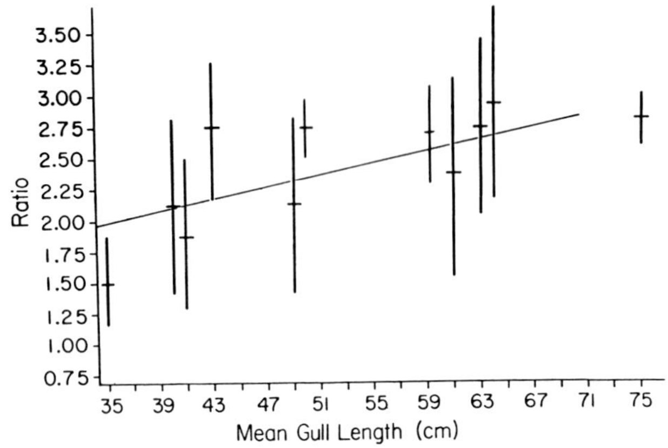

FIGURE 5. Foraging ratio (young interfood interval/adult interfood interval) as a function of age of maturity.

sites with both foraging adults and young, I eliminated those foraging sites with long interfood intervals for adults, but where young were unable to feed. In some cases, one or two young were present but they obtained no food.

Secondly, I examined only foraging situations where there were 20 or more of both adults and young. In foraging sites where there were many adults but only a few young, the young were usually completely unsuccessful at obtaining food items or required over three times the time adults required to find food items (unpublished data). Thus, I did not include in the data set foraging situations with very large differences in interfood intervals between adults and young. Had I included these two types of foraging sites, the disparity between adults and young would have been even larger than those I report, so it would strengthen the differences rather than weaken them. They were not included because it was impossible to obtain sufficient sample sizes of young for statistical analysis.

By using interfood interval, I have chosen not to examine habitat location or selection. Both aspects of foraging behavior are extremely important, and young are presumably less able to locate feeding sites, to decide whether to forage there, and to decide when to abandon a site in search of a more productive one.

Interfood Interval and Age

For over 90% of the data sets examined in this study adults had significantly shorter interfood intervals than young. Thus, assuming that they are capturing the same size prey within any given feeding situation, adults are obtaining more food per unit time than young. Comparable results have been reported for several other seabirds, but this report illustrates the consistency of this finding for a large number of congeners feeding in very different habitats on a variety of food items.

The interfood interval includes search, capture, and manipulation time since all three activities occur between successfuly capturing a first and second food item. From the results of this study it is not possible to determine which of these three (or combination thereof) is difficult for the young gulls. The literature (Verbeek 1977, Ingolfsson and Estrella 1978, Gochfeld and Burger 1981, MacLean 1986), however, suggests that across species young birds may be less efficient at all three tasks.

Ratio of Young to Adult Interfood Interval and Foraging Task Difficulty

I had originally predicted that the ratio of young to adult interfood interval would increase with the difficulty of the task as measured by adult interfood interval. This turned out not to be the case. This can be accounted for by two factors: (1) adult interfood interval may not be the best measure of foraging difficulty and (2) young gulls actively avoid difficult foraging situations, resulting in only those young gulls that are proficient in foraging in these sites. Adult interval may not be the best basis to assess foraging difficulties if in short interfood interval situations young have difficulty maneuvering prey. If adult interval is only 2 sec, and young require 2 sec longer to handle food, then they are twice as slow as adults whereas when adult interfood interval is 120 sec an additional 2 sec by young in food handling does not appreciably increase young interfood interval.

Young gulls clearly avoid the most difficult foraging tasks (i.e., plunge-diving) and it was often difficult to find enough young for a sufficient sample. Thus perhaps only the young already proficient at difficult tasks choose to forage there, minimizing the differences between adult and young intervals. Alternatively, since all young must feed somewhere, the less proficient young might concentrate in relatively easy foraging situations, maximizing the differences between adult and young interfood intervals.

Interfood Interval, Gull Size and Age of Maturity

One factor that supposely contributed to the evolution of delayed maturity in seabirds is the difficulty young and subadults have in foraging efficiently (e.g., Lack 1967, Ashmole 1971). This assumption can be indirectly tested by examining the disparity between the foraging efficiency of adults and young of closely related species that mature at different ages. An important purpose in this study was to examine the disparity in *Larus* gulls to determine if the disparity is greater in gull species with greater delayed maturity. Data from 106 foraging sets with 15 species of gulls maturing from two years to five years of age confirmed the prediction that disparity is greater for gulls that mature later. Age of maturity and size are directly related, and in this study the disparity was also directly related to size of the gull species. These results are clearly consistent with the theory that foraging difficulties contributed to the evolution of delayed maturity in gulls.

ACKNOWLEDGMENTS

I gratefully acknowledge the help of Michael Gochfeld with logistics, data collection and analysis, and for comments on the manuscript. I thank David and Deborah Gochfeld for help in data collection, D. Sonneborn for logistical support in Alaska, and B. Lauro and R. Steidl for data analysis.

This work was supported in part by Ehrenkrantz Associates and Rutgers University. Manuscript preparation was aided by funds from the Charles and Johanna Busch Fund of Rutgers University.

LITERATURE CITED

Ainley, D. G., E. F. O'Connor, and R. J. Boekelheide. 1984. The marine ecology of birds in the Ross Sea, Antarctica. Ornithological Monograph No. 32, Washington, D.C.

Ashmole, N. P. 1963. The regulation of numbers of tropical oceanic birds. Ibis 103b:458–473.

Ashmole, N. P. 1971. Seabird ecology and the marine environment. Pp. 223–286 *in* D. S. Farner and J. B. King (eds.). Avian biology. Vol. 1. Academic Press, New York.

Ashmole, N. P., and M. J. Ashmole. 1967. Comparative feeding ecology of seabirds of a tropical oceanic island. Peabody Mus. Nat. Hist. Bull. 24:1–131.

Buckley, F. G., and P. A. Buckley. 1974. Comparative feeding ecology of wintering adult and juvenile Royal Terns (Aves: Laridae: Sterninae). Ecology 55:1053–1063.

Burger, J., and M. Gochfeld. 1981. Age-related differences in piracy behaviour of four species of gulls, *Larus*. Behaviour 77:242–267.

Burger, J., and M. Gochfeld. 1983. Feeding behavior in Laughing Gulls, compensatory site selection by young. Condor 85:467–473.

Burger, J., J. R. Trout, W. Wander, and G. S. Ritter. 1984. Jamaica Bay Studies VII: factors affecting the distribution and abundance of ducks in a New York estuary. Estuarine, Coastal and Shelf Science 19:673–689.

Diamond, A. W. 1984. Feeding overlap in some tropical and temperate seabird communities. Pp. 24–46 *in* R. W. Schreiber (ed.). Tropical seabird biology. Studies in Avian Biology No. 8.

Duffy, D. C. 1983. The foraging ecology of Peruvian seabirds. Auk 100:800–810.

Dunn, E. K. 1972. Effect of age on the fishing ability of Sandwich Terns *Sterna sandvicensis*. Ibis 114: 360–366.

Dunnet, G. M., and J. C. Ollason. 1982. The feeding dispersal of Fulmars *Fulmarus glacialis* in the breeding season. Ibis 124:359–361.

Dwight, J. D. 1925. The gulls (Laridae) of the world: their plumages, moults, variations, relationships and distribution. Bull. Am. Mus. Nat. Hist. 52:63–401.

Gochfeld, M., and J. Burger. 1981. Age-related differences in piracy of frigatebirds on Laughing Gulls. Condor 83:79–82.

Hunt, G. L., Jr., and M. W. Hunt. 1973. Habitat partitioning by foraging gulls in Maine and Northwestern Europe. Auk 90:827–839.

Ingolfsson, A., and J. J. Estrella. 1978. The development of shell-cracking behavior in Herring Gulls. Auk 95:577–579.

Krebs, J. R., A. L. Houston, and E. L. Charnov. 1981. Some recent developments in optimal foraging. Pp. 3–18 *in* A. C. Kamil and T. D. Sargent (eds.). Foraging behavior: ecological, ethological and psychological approaches. Garland STPM Press, New York.

Lack, D. 1967. Interrelationships in breeding adaptations as shown by marine birds. Proc. 14th Int. Ornithol. Congr. 3:42.

MacLean, A. A. E. 1986. Age-specific foraging ability on the evolution of deferred breeding in three species of gulls. Wilson Bull. 98:267–279.

Morony, J. J., Jr., W. J. Bock, and J. Farrand, Jr. 1975. Reference list of the birds of the world. American Museum of Natural History, N.Y.

Morrison, M. L., R. D. Slack, and E. Shanley, Jr. 1978. Age and foraging ability relationships of Olivaceous Cormorants. Wilson Bull. 90:414–422.

Mudge, G. P., and P. N. Ferns. 1982. The feeding ecology of five species of gulls (Aves: Larini) in the inner Bristol Channel. J. Zool. Lond. 197:497–510.

Orians, G. H. 1969. Age and hunting success in the Brown Pelican *Pelecanus occidentalis*. Anim. Behav. 17:316–319.

Ryder, J. P. 1980. The influence of age on the breeding biology of colonially nesting seabirds. Pp. 153–168 *in* J. Burger, B. L. Olla, and H. E. Winn (eds.). Behavior of marine animals, Vol. 4 Marine birds. Plenum Press, New York.

SCHOENER, T. W. 1971. Theory of feeding strategies. Ann. Re. Ecol. and System. 2:369–404.

SEARCY, W. A. 1978. Foraging success in three age classes of Glaucous-winged Gulls. Auk 95:587–588.

ULFSTRAND, S. 1979. Age and plumage associated differences of behavior among Black-headed Gulls *Larus ridibundus*: foraging success, conflict, victoriousness and reaction to disturbance. Oikos 33:160–166.

VERBEEK, N. A. M. 1977. Age differences in the digging frequency of Herring Gulls on a dump. Condor 79:123–125.

Studies in Avian Biology No. 10:91–101, 1987.

FORAGING PATTERNS AND PREY SELECTION BY AVIAN PREDATORS: A COMPARATIVE STUDY IN TWO COLONIES OF CALIFORNIA GULLS

JOSEPH R. JEHL, JR.[1] AND CHARLES CHASE III[2]

ABSTRACT.—We studied the effects of avian predators, especially Great Horned Owls (*Bubo virginianus*), in colonies of California Gulls (*Larus californicus*) at Mono Lake, California, and Antero Reservoir, Colorado, in the years 1981–1985. During early gull nesting and incubation owls preyed on adult gulls. The frequency of kills varied from almost nightly in the Mono Lake study area in some years to as little as monthly at Antero Reservoir. Owls shifted to chicks as they became available. During attacks adult gulls left the colony for periods of 30 min to 3 h. This resulted in indirect chick losses caused by hypothermia at Antero Reservoir, where nocturnal temperatures often dropped below 0°C, and by subsequent predation by adult gulls at Mono Lake, where temperatures were moderate.

Losses of adult gulls to owls were negligible, whereas losses of chicks were sometimes great, and in some areas nesting was completely disrupted. Predation by Golden Eagles (*Aquila chrysaetos*) and Common Ravens (*Corvus corax*) was infrequent and had little effect on adults or young at either colony.

We also investigated factors that make gulls susceptible to avian predators. Adult size was unimportant, but females comprised 70% of the kills, probably owing to their greater attentiveness at the nest. Gulls nesting near the owls' nest or in areas in which predation had been high in previous years also risked greater mortality. Gulls in dense, central parts of the colony suffered very little mortality at Mono Lake, but high mortality at Antero Reservoir. Nesting near vegetation greatly increased the risk of owl predation in some areas; in others it had little effect. However, nesting where visibility was obstructed by plants or topography facilitated attacks by Golden Eagles.

Gull distribution within the colony changed in years following intense owl predation. Experienced birds shifted to new nest sites or left the area entirely and were replaced by birds that had nested in peripheral areas or by birds nesting for the first time.

Knowledge of the impact and history of predation in a colony is relevant to interpreting patterns of colony occupancy and habitat use.

Great Horned Owls (*Bubo virginianus*) are important predators on colonially nesting larids (Southern et al. 1982, Nisbet 1975), including California Gulls (*Larus californicus*) (Vermeer 1970). Not only do they kill adults and young, their nocturnal raids can disrupt colonies, causing adults to desert eggs and chicks (Nisbet and Welton 1984). In studies at Mono Lake, Mono Co., California, Jehl observed that these owls as well as Golden Eagles (*Aquila chrysaetos*) hunted regularly in the large California Gull colony, sometimes having a major impact on nesting success. Concurrently, Chase obtained similar results from studies at Antero Reservoir, Park Co., Colorado, where owls, eagles, and Common Ravens (*Corvus corax*) were present. In this paper, we (1) describe the behavior of predators and prey, which varied annually both within and between the two colonies, (2) compare the impact of nocturnal vs. diurnal predators on the gulls' reproductive success, (3) present information on prey selection by the owls and eagles, and (4) document distributional changes in the colonies resulting from these disturbances.

[1] Sea World Research Institute/Hubbs Marine Research Center, 1700 South Shores Rd., San Diego, California 92109.

[2] 1810 Belaire, Denver, Colorado 80220.

METHODS

MONO LAKE, CALIFORNIA

From 1982 through 1985 Jehl studied California Gulls breeding on the Paoha and Channel islets (Fig. 1A) at Mono Lake (ML) (elevation ca. 1945 m). In 1982, these comprised ca. 15 sandy and virtually barren islets (Paoha Islets) plus many tiny satellites, and two rocky islets (Channel Islets) approximately 0.6 km to the NE. They varied in size from about 0.02 to 6 ha. In 1983, a rise in lake level and severe erosion greatly reduced the number and extent of the islets, so that in 1984–1985 only five islets and their satellites persisted in the Paoha series (Fig. 1B); the Channel Islets were submerged. In the years 1982–1985 the nesting population on these islets varied from 6200 to 17,000 gulls (14–35% of the entire ML colony).

Laying at ML begins in late April, peaks between 10–30 May, and continues into mid-June. The hatching period peaks in the first two weeks of June. In May nocturnal temperatures occasionally fall below freezing, but later in the summer, when chicks are present, they rarely dip below 10°C. Diurnal temperatures in the chick-rearing period commonly exceed 30°C.

Predation was studied opportunistically in 1982–1983 and more systematically in 1984–1985. In the latter two years, Jehl surveyed the colony regularly from the start of the breeding season in early April until all fledglings had departed in August. Because the nesting islets are low (maximum elevation 0.3–2.5 m in 1984), flat, and essentially unvegetated (Fig. 2), he could detect most dead adults from a small boat and landed only momentarily to retrieve carcasses. To determine mor-

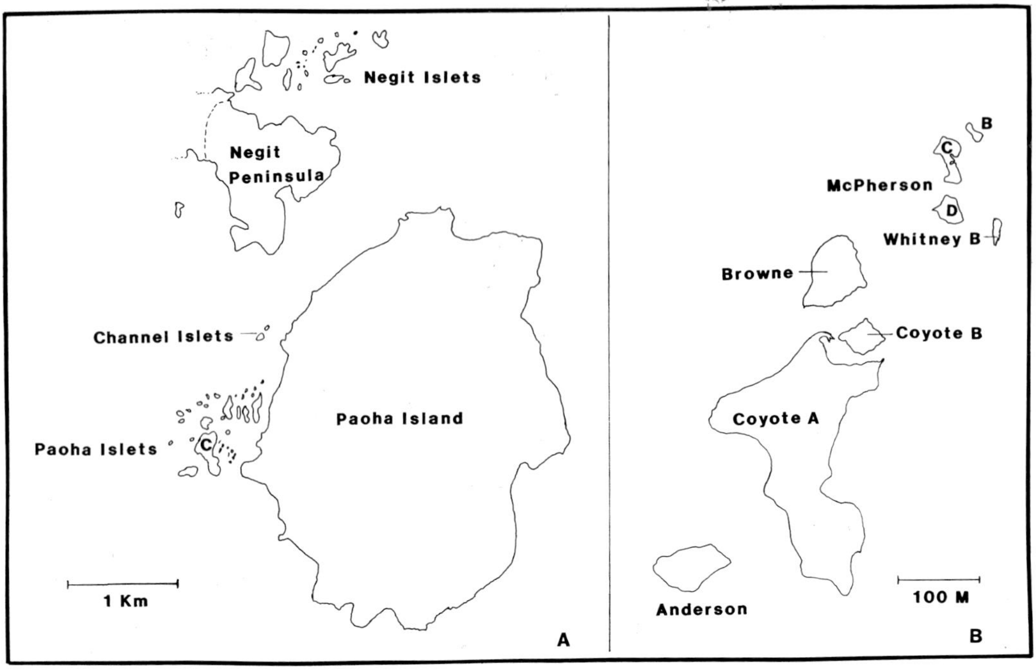

FIGURE 1. A) Configuration of the islands in Mono Lake, California, in 1982; C = Coyote Islet. B) Configuration of the Paoha Islets in 1984–1985.

tality among young gulls, he searched each islet by foot on 9–10 July, when the earliest chicks were fledging; in late July, when most young were independent; and in early August, after the islets were deserted. Other checks were made opportunistically in areas where nesting had failed, or where the brief presence of an observer would not affect chick survival.

Jehl removed all carcasses, plotted the location of those that had been killed by predators and, when possible, obtained standard morphological measurements (exposed culmen, depth of bill at gonydeal angle, chord of wing, tarsus) on remains of adults. Adults were sexed by gonadal examination if possible, or by measurements. Size differences allow over 90% of adults to be sexed correctly (Jehl 1987).

Antero Reservoir, Colorado.—California Gulls nest on three islands in Antero Reservoir (AR) (elevation 2770 m). These are small (0.5 to 17 ha) and moderately vegetated with grasses and low sage shrubs. This colony, which has existed for at least 30 years, increased from 643 pairs in 1981 to 1218 in 1985. Breeding chronology is similar to that at ML, although a late thaw may delay the season by 1–2 weeks. Temperatures regularly fall below freezing through the first two weeks of brooding. Snow, up to 0.2 m, is not uncommon through mid-June. High diurnal temperatures of 21–30°C in June and July are regularly reduced by daily thunderstorms. Temperature variations may exceed 22°C per hour and 30°C per day.

Chase documented the breeding on Rock and Gull islands (Fig. 3), at the western end of the reservoir and on Goose Island, at the eastern end, since 1981. He made daily visits to each island and documented mortality through the entire breeding season. From 20 to 40% of the adults in the colony were individually marked and their breeding efforts were monitored each year.

RESULTS

Great Horned Owl

Foraging patterns

The presence of Great Horned Owls in each colony was ascertained by direct observations and could also be inferred from remains at the kill site. Owls typically dismember gulls, biting off the head, and sometimes wings and legs, and feeding mainly on the upper breast (Fig. 4). They fed at or near the point of capture, and removed very few gulls, as confirmed by surveys of their nests and roosts. Jehl twice found decapitated adults (wt. ca. 600 g) floating 100 m offshore the colony, suggesting that these were too heavy to be carried off. From these observations we think that owl foraging areas and impact on adult gulls and large chicks can be measured satisfactorily. Small chicks disappear for many reasons, including predation by adult gulls, and their losses to owls are often unmeasurable.

One or two owls foraged regularly in the study area at ML. At AR predation was infrequent and no more than one owl was present per night. The differences may reflect the existence of alternative food sources; they were not related to distance from the owls' nest or roost (2–2.5 km from the foraging areas in each case). At both localities

FIGURE 2. Typical nesting area of California Gulls on Paoha Islets, Mono Lake, California.

owls began capturing adults shortly after the gulls initiated strong territorial defense in mid-April, and continued taking them through mid-June, when they switched to chicks. Over 90% of adult kills occurred from late April to early June, predation being heaviest in the last half of May (Fig. 5). At AR an owl would fly in and make a direct kill of an adult. When preying on chicks, it would typically land on the crest of a hill and, after scaring off adults, fly into a group of chicks. We have no observations of hunting behavior at ML.

In each year at ML the owls maintained consistent and rather localized hunting ranges. They appeared to start hunting near their nest site, almost surely on Negit Island, and to work over a small area until its resources were depleted. In 1982, predation was first noted on the Channel Islets, which were deserted by 6 July after several large chicks and adults were killed. The owls then shifted to Coyote Islet and later in the season to McPherson Islet. In June 1983, although a few chicks were killed early on Coyote, the owls evidently concentrated their activity on the Channel Islets, routing all 81 pairs by 10 July. They then shifted to Whitney B; all 23 pairs deserted within three days. Subsequently, as in 1982, they hunted on Coyote and then on McPherson in late July and early August. They also continued to forage on Whitney B, killing juveniles that had arrived from other islets.

In 1984, the Channel Islets were submerged.

FIGURE 3. Nesting area on Gull Island, Antero Reservoir, Colorado.

FIGURE 4. The decapitated remains of a California Gull killed by a Great Horned Owl on Negit Island, Mono Lake, California, 1985.

The earliest kills were on Browne and Coyote B islets, but from mid-June through mid-July owls hunted exclusively—and totally disrupted nesting—on the southern end of Coyote A. Subsequently, they moved to the north end of Coyote A, hunting there and on Browne Islet until mid August (Fig. 6).

In 1985, as in 1984, hunting was concentrated on Coyote A Islet (Table 1), but was less intense and more dispersed. It was first noted on 16 June, when three chicks were killed on the southern tip of the islet. Except for four large chicks killed in mid-July, there was no further evidence of predation until late July, when large chicks were taken nightly at least through 8 August, mostly on the northern half of the island.

At AR predation was confined to Gull Island until 1984, when a large aggregation of gulls moved to Goose Island (Table 1). The owl shifted its activities among areas of high nesting density on Goose Island.

In the ML study area, virtually all adults and young were dispatched in open areas on relatively high ground. However, when ca. 15 pairs began nesting among dense shrubbery near the owl's roost on Negit Island in 1985, predation was far greater than in any other area or year; at least nine adults were killed and the remainder abandoned before any eggs hatched. At AR adults were also killed on high bare ground, whereas chicks were taken without regard for topography but always in areas of high density. In 1982 mortality was greatest in vegetation near the tip of Gull Island, whereas in 1983 it was greatest in the barren center of the island.

At ML owls avoided dense concentrations of gulls, foraged on the periphery of nesting areas, and did not shift into areas of high density until late in the season, after most young had fledged and birds were few. By contrast, at AR central nests attracted more attention.

Throughout this study there was no evidence of owl predation away from the AR colony. At ML, thousands of gulls gather daily on the deltas of small creeks. Sick or weak birds that may remain ashore there at night are killed by owls (distinct from those hunting in the colony) and coyotes (*Canis latrans*), but the numbers involved are small.

Impact

In both colonies, the impact on adult gulls was small, amounting to 0.2–1.5% (average 0.4%) of the nesting population (Table 1); nearly all mortality occurred before mid-June. At ML in 1984 observations began on 5 April but predation did not begin until mid-April (Fig. 5), presumably because gulls left the islands at night prior to the start of intense territorial defense (Vermeer 1970, Chardine and Morris 1983). Over the next 60 days 60 adults were killed (1.0/night). Mortality was not constant, however, and in late May 30

TABLE 1

NUMBERS OF ADULT GULLS KILLED BY OWLS ON THE PAOHA ISLETS, MONO LAKE, CALIFORNIA, AND ANTERO RESERVOIR, COLORADO, 1981–1985

	1981		1982		1983		1984		1985	
	No. nesting	No. killed	No. nesting	No. killed	No. nesting	No. killed	No. nesting	No. killed[a]	No. nesting	No. killed[b]
Paoha Islets										
Islets										
Anderson							426	0	394	0
Browne							1936	6 (0.3%)	1134	2 (0.2%)
Coyote A							2288	56 (2.4%)	2492	22 (0.9%)
Coyote B							446	2 (0.4%)	332	0
McPherson B							68	0	58	0
McPherson C							970	0	1076	0
McPherson D							956	0	810	0
Whitney B							6	0	6	0
Total							7096	64 (0.9%)	6302	24 (0.4%)
Antero Reservoir										
Gull	550	10 (1.8%)	584	5 (0.8%)	450	1 (0.2%)	416	3 (0.7%)	466	0
Goose	50	0	75	0	134	0	496	1 (0.2%)	715	5 (0.7%)
Rock	43	0	39	0	32	0	36	0	37	0
Total	643	10 (1.5%)	798	5 (0.6%)	616	1 (0.2%)	948	4 (0.4%)	1218	5 (0.4%)

[a] 14 April–31 August; one additional adult found killed on 7 September 1984.
[b] 4 April–17 August; one additional adult found killed on 25 September 1985.

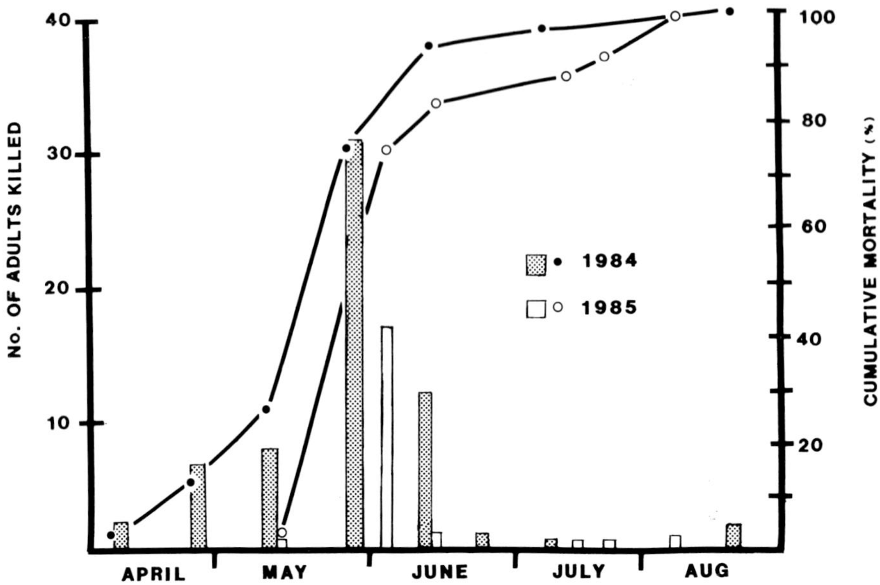

FIGURE 5. Mortality patterns of adult California Gulls killed by Great Horned Owls on the Paoha Islets, Mono Lake, California, 1984 and 1985.

adults were killed in 15 days. In 1985 owls did not appear in the study area until 14 May, and from then until 4 June 18 adults were killed (0.86/night); only six other adult kills were recorded subequently. At AR only 26 adults were killed over five nesting seasons, all in the incubation stage.

Females suffered most of the adult losses in 1984–1985 at ML (no data for other years). Of 112 killed, 78 were females (69.6%), 18 (16.1%) males and 16 (14.3%) unsexed. This differs significantly from a 1:1 sex ratio and also from the sex ratio of gulls found dead in the colony of other causes (56 females : 42 males; $P < 0.005$ in each case, χ^2 test). The sex ratio of kills made away from the colony was: males—4, females—2, unsexed—1. Dimensions of gulls killed by owls did not differ from those in the population at large (Student's t-test; Table 2).

Losses of young were higher than of adults, and much more difficult to assess because (1) small chicks disappear without a trace (some are lost to predators, and others are devoured by adult gulls); and (2) indirect effects of predation can greatly exceed those from killing, and can only be inferred.

At ML in 1982, no chicks were produced from 20 nests on the Channel Islets, and the number of young killed on the Paoha Islets (78) amounted to 2.8% of the total alive in early July; at least 33 chicks were killed in a 13-day period in late July–early August. In 1983, predation again resulted in complete failure on the Channel Islets (81 nests) and Whitney B (23 nests) and resulted in an estimated loss of 2–3% of the chicks in the study area. The intensity of predation in late July–early August (31 kills in 14 days) was similar to that in 1983.

In 1984, direct mortality, although higher than in other years, remained low. An estimated two young were killed each night from mid-June to mid-August (120 chicks). Indirect mortality, however, was great. No chicks are known to have fledged from 377 nests on the southern end of Coyote Island, the owl's main foraging area (Fig. 6). This was 33% of the nests on that island and 10.6% of those in the study area. Further, if each of 52 adults killed prior to 23 June resulted in the loss of a nest, nest failure from owls rises to at least 12%.

TABLE 2
MEAN DIMENSIONS (MM) OF ADULT CALIFORNIA GULLS AT MONO LAKE, CALIFORNIA

	Killed by Great Horned Owls			Collected or found dead		
	Sex	N	Mean	Sex	N	Mean
Exposed culmen	M	5	47.3	M	45	47.2
	F	17	42.9	F	85	42.7
Wing (chord)	M	6	389.8	M	47	393.8
	F	34	372.5	F	94	374.5
Tarsus	M	0		M	42	59.2
	F	10	54.0	F	82	55.1

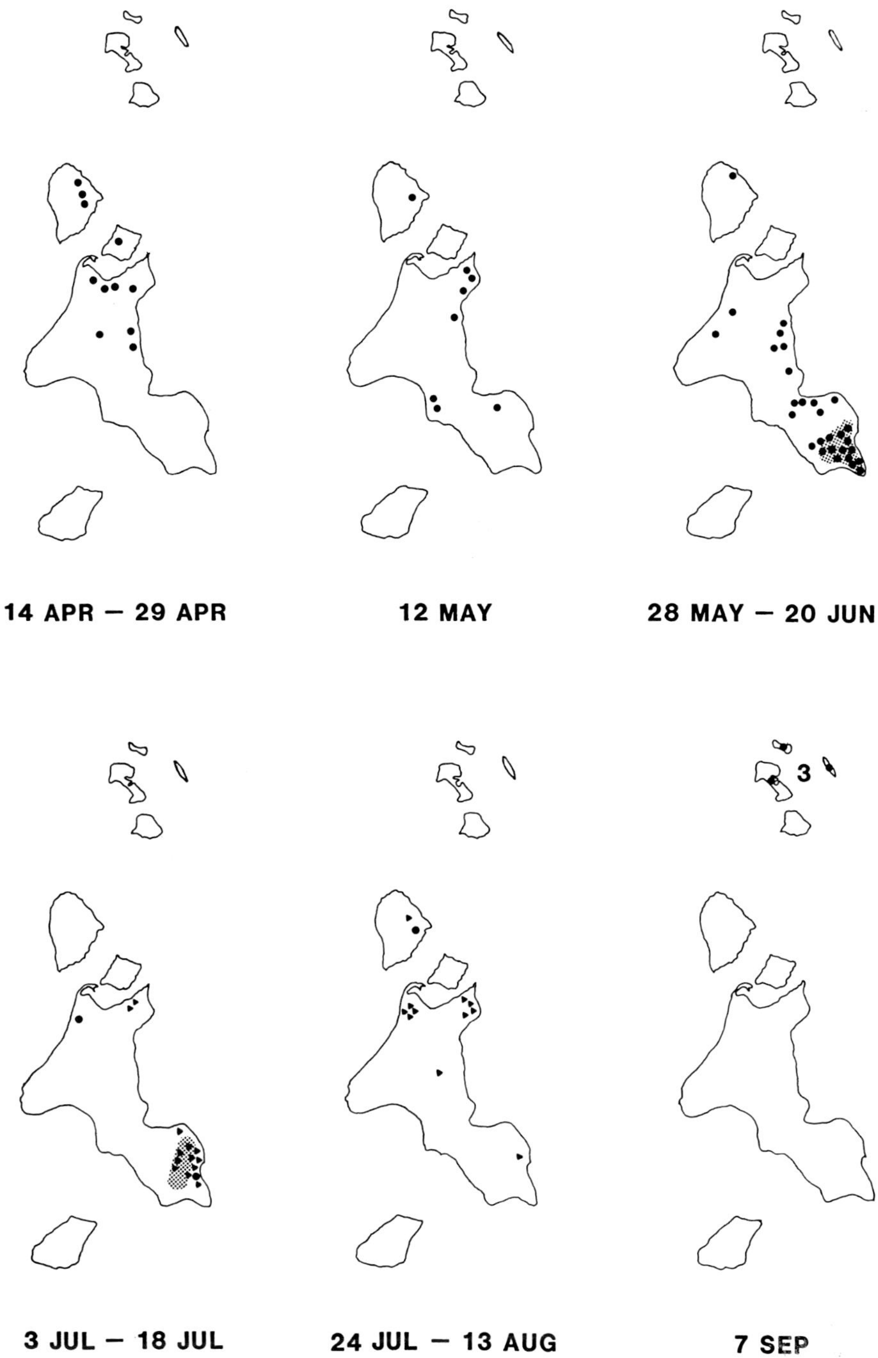

FIGURE 6. Hunting locations of Great Horned Owls on the Paoha Islets, Mono Lake, California, in 1984. Circles indicate kills of adults, triangles of young gulls. Stippled area shows major hunting area for young on Coyote A; some kills in these areas are not plotted individually.

TABLE 3
DIRECT AND INDIRECT MORTALITY OF CHICKS FROM OWL PREDATION AT ANTERO RESERVOIR, COLORADO

Year	Direct	Indirect
1981	15–20 (ca. 50%)	16 (50%)
1982	74 (72.5%)	28 (27.5%)
1983	18 (22.2%)	63 (77.7%)
1984	45 (46.5%)	54 (53.5%)
1985	38 (29.9%)	89 (70.1%)

At AR owls caused the death of 3–10% of chicks in any year. Mass killings sometimes occurred when an owl entered a dense area of the colony; although only one or two chicks were eaten, many might be dismembered (51 in 30 minutes by one owl on 21 June 1982). Nevertheless, indirect mortality sometimes exceeded direct mortality (Table 3). It resulted mainly from exposure and occurred even though chicks formed creches. Temperatures below 0°C, when coupled with rain, resulted in death from hypothermia for untended chicks (e.g., 16 on 20 June 1981). The chicks' tendency to creche when the adults were driven off led to further local increases in density, which facilitated predation.

Responses to owls

We have no direct observations from ML, although nocturnal desertion of gull nests has been noted (M. Morton pers. comm.). At AR adults left the colony as soon as a kill was made (which was as soon as the owl arrived), but usually returned within 30 min after the owl departed. On nights of intense predation, eggs and chicks might remain unattended for several hours, but adults never remained away until daylight (cf. Southern et al. 1985). In areas of persistent and localized predation at ML, adults often abandoned eggs and chicks. This did not occur at AR, where the owl's activities were more widespread.

The effects of predation extended over more than a single year and affected the use of breeding sites in subsequent seasons. At AR in 1981, predation was insignificant, and colony structure and distribution remained stable into 1982. In 1982–1984, predation became important on Gull Island, and adults began to shift to other areas on Gull Island and invaded Goose Island in 1983. By 1985 many birds had moved to Goose Island; many others, however, never reappeared and presumably joined colonies that were forming elsewhere in the western Great Plains (Chase pers. obs., Findholt 1986). Data from marked adults showed that experienced breeders were involved in shifts within the colony, and that their previous nest sites were claimed by 2–4-year-old birds new to the area, or by 4–5 year olds that had nested on the periphery of the colony before (Chase in prep.). Local changes in distribution also occurred at ML; in the most dramatic case, an area with persistent predation in 1984 was occupied two weeks later than other areas in 1985, and the number of pairs there dropped from 377 to 109.

GOLDEN EAGLE

Foraging patterns

Eagles usually leave carcasses intact but eat the entire pectoral muscle mass as well as the neck and major limb muscles. Unlike owls, they can easily carry off adult gulls (Behle 1958). One or two eagles hunted regularly at ML, albeit infrequently in the study area. Kills were recorded from 24 April to 28 July; most mortality occurred by mid-June and, accordingly, was concentrated on adults. Forty of 41 kills in the study area took place on Coyote A and were made in depressions that contained some vegetation.

Impact

In all years, at both areas, the impact of eagles was small and probably restricted to direct mortality, as their disturbances were diurnal and too brief to result in the redistribution or exposure of chicks. At ML from 1983–1985, 32 adult and nine juvenile gull carcasses were attributed to eagle predation, 25 in 1983 alone. No kills were recognized in 1982, but the data are incomplete. As with owls, most eagle-caused mortality among adults was suffered by females (16 of 21 sexed kills), which did not differ in size from females in the colony at large (Jehl unpubl.). At AR eagles killed a total of three gulls in 1984–1985.

Responses to eagles

The arrival of eagles at ML caused a brief panic flight among gulls, even when the predators were as much as two km overhead. At AR eagles cannot approach undetected and were usually attacked if they approached within 0.5 km of the colonies. Indeed, one was driven into the lake and, after swimming ashore, was harried constantly as it dried off.

Jehl observed two hunting episodes. In one, an eagle flew low across ML from the southwest, intersecting Coyote Islet in an area where few gulls were nesting. Using the crest of the islet for concealment, it cruised, accipiter-like, 0.5 m above the ground, then veered sharply across the ridgeline and plunged into a hollow, which was found to contain the remains of 11 adult gulls. Nesting gulls panicked, but settled within 2 m of the eagle after it landed. When it flushed, a few gulls mobbed it briefly but most continued incubating. In another event, two eagles flushed from an area of scattered brush on Paoha Island, where they had been chasing a sick gull. Although no more than 50 m from a group of loafing gulls, the gulls showed no reaction until the eagles flushed.

Other Predators

At AR Common Ravens were harried when they approached and could capture chicks on the periphery of the colony only when adults were distracted (e.g., by the presence of humans). They accounted for the loss of 16 small (<200 g) chicks, but no eggs. Canada Geese (*Branta canadensis*) defending nests killed two adults and one chick that strayed nearby. A muskrat (*Ondatra zibethica*) entered the colony in 1981 and ate five clutches before it was attached and killed by adult gulls.

DISCUSSION

Southern et al. (1982) noted that "the following generalizations seem to apply with respect to nocturnal predation on colonial gulls: (1) adult gulls usually react . . . by temporarily fleeing the site, (2) fleeing exposes eggs and young to . . . weather and predation, and (3) no aspect of gull breeding biology . . . appears to be effective in decreasing the toll. Additionally, tenacity to a colony site reduced the probability that a gull will relocate in successive years following nesting failures." Our data support points 1 through 3, but, we found that site tenacity was relinquished in years following intense predation. Studies of marked gulls at AR showed that stability was an illusion, which resulted from the replacement of previous inhabitants by new birds.

At both colonies indirect losses after periods of nocturnal disturbance exerted a higher toll on chicks than did outright killing, but the causes differed. Death at ML resulted from predation by adult gulls, whereas at AR hypothermia was the major cause. Chardine and Morris (1983) also found that "the cost of desertion at night, in terms of egg failure, was relatively small" when avian predators were involved. Much higher losses have been reported in response to mammalian disturbance (Southern et al. 1985).

The owls' exploitation patterns also differed at the two colonies. At ML they hunted rather predictably, slowly depleting prey in one area before moving on. At AR the owl hop-scotched between areas of high abundance and sometimes indulged in frenzy killing. As a result, the immediate impact at AR was greater, and this may have made it disadvantageous for the owl to return to an area visited previously. These exploitation patterns may also have affected the responses of adult gulls. At ML owls caused relatively little chick mortality on any night, but their repeated visits caused adults to abandon eggs and chicks (cf. Emlen et al. 1966, Vermeer 1970). At AR, episodes of predation, though locally severe, were non-recurring and parents remained as long as their chicks survived.

The Basis of Prey Selection

Which factors determine whether individuals will fall prey to predators? Studies of diurnal predators have indicated that very young or very old individuals, as well as those that are ill, out of range, behaving unusually, or are oddly colored are prone to high mortality (Rudebeck 1950–1951, Lack 1954, Mech 1970, Mueller 1974). How nocturnal predators select prey has received little attention, although several studies have indicated that owls are likely to catch individuals that are inexperienced or in unfamiliar terrain (e.g., Pearson and Pearson 1947, Metzgar 1967).

Prey selection by owls

Owls probably hunted entirely by sight in our colonies. Even on the darkest clear nights there was sufficient starlight to highlight roosting or incubating gulls, whose abundance made hunting by aural cues unnecessary. Owls at the Paoha Islets evidently originated on Negit Island. Because owls tended to begin foraging near their own nest before moving farther afield, nearby gulls suffered high mortality. Data showing that owls returned annually, in a regular sequence when possible, to areas that had been bountiful in the past, indicate that local knowledge plays a major role in hunting success, even when food is superabundant and conspicuous (cf. Martin 1986).

We found, as did Vermeer (1970), that owls preyed on adults early in the season but then shifted to young. The switch might indicate a preference for chicks, but more likely indicates the chicks' greater vulnerability. Adults, having no defense against nocturnal predators, have no alternative but to depart. At AR owls in a feeding frenzy took only large young, which ran and creched; they ignored nearby small chicks, which usually crouched motionless, a behavior that rendered them susceptible to exposure.

The position of a nest in a colony affects its chances of being raided by a diurnal predator; peripheral sites are more vulnerable because the predator has less chance of being detected and endures fewer attacks to reach them (Buckley and Buckley 1980). Nest position would seem of little relevance to owls, because gulls neither mob nor defend their nest against them (Kruuk 1964, 1966). Nevertheless, owls at ML avoided the center of the colony and hunted there only late in the season, when densities had become very low. This suggests that some aspect of population size or density can impair the success of nocturnal predators as it does for diurnal predators (Burger and Gochfeld 1984). Yet, owls at AR foraged mostly where densities were greatest, and central nests attracted most attention. These differences may be related to the ease with which prey can be detected. At ML the substrate is relatively uniform, and prey are easily perceived. At AR scattered vegetation provided greater camouflage for the chicks, which may have caused the owls to hunt where chicks were most abundant.

At AR and on the Paoha Islets, many kills of adult gulls took place on high, bare ground, evidently because birds there were more quickly detected by arriving owls. We found no consistent relationship between the presence of vegetation and chick mortality at AR; vegetation on the Paoha Islets is too scarce to allow any conclusions. However, in 1985, when gulls nested among dense shrubbery on Negit Island, adults suffered high losses. By nesting in that habitat the gulls achieved concealment but, when discovered, found their escape routes blocked by plants. Thus, it is not surprising that in all locations at ML vegetated areas were consistently the last to be occupied (Jehl and Mahoney MS).

Adult females were highly susceptible to aerial predators (68% of owl kills; 76% of losses to eagles), probably owing to their greater attentiveness. Jehl and Mahoney (1983) inferred differences in the time budgets of male and female California Gulls, and Chase (unpubl.) found that 73–89% of nests were brooded by females from 2200–0400 hrs, and that males often roosted off the territory. Although females average 20% lighter than males, size did not affect an adult's vulnerability because gulls of either sex killed by owls were no smaller than those in the general population, and because even males are far smaller than the maximum size prey for Great Horned Owls. We know of no previous demonstration of differential mortality from avian predators in gull colonies.

Prey selection by eagles

Eagles are visual predators, and at ML captured prey almost exclusively in two small areas near the edge of Coyote Islet. Although our observations are consistent with the ideas that familiarity with an area is important, and that peripheral nests are more likely to be attacked, physiographic conditions seemed to play the overriding role in hunting success. All but one kill was made where vegetation or undulating terrain impaired the gulls' ability to detect arriving eagles.

Consequences of predation

Sargeant et al. (1984) reported that red foxes (*Vulpes vulpes*) (1) killed a total of 242,000 adult ducks each year in the Prairie Potholes region of North Dakota, (2) 76% of the kills in the north-central United States were females, and (3) anatid populations are highly skewed toward males (1.2:1). In our study the number of adult gulls killed by avian predators was too small to have affected the operational sex ratio, and the number of young killed, directly or indirectly, was too small to exert a major influence on population dynamics, although local effects might be extreme. If predators imposed any significant selection on nesting gulls, it is likely to have been related to promoting synchrony and affecting the choice of nesting habitats. Early-nesting adults and late-fledging chicks have a high probability of being captured, because they are the only prey available. This is also true for adults that nest in areas previously frequented by predators or in habitats where the ability to detect or flee from arriving predators is impaired.

Measuring mortality

Predation is common in seabird colonies. Its direct effects are obvious, but its indirect effects can be hard to perceive and quantify. For example, in 1984, when gull nesting density was greater than in other years on the Paoha Islets, the effects of owls were greater and those of eagles less than in other years. We suspect that these events were interrelated; specifically, that the high density of gulls enhanced indirect losses to owls, but concomitantly made it nearly impossible for eagles to enter the study area undetected during the day.

In studying colonial birds, researchers often determine life history parameters in a "typical" area and extrapolate the results to the entire colony, even though it is impossible to judge *a priori* which study areas, if any, might provide "representative" results. As we have shown, annual variability in the local impact of predators can be large and lead to the abandonment of long-held nesting areas by experienced birds and their later occupancy by inexperienced breeders, which in turn can result in changes in the age structure of a breeding unit (Pugesek and Diem 1983) and in seemingly inexplicable changes in productivity (Chase in prep.). Students of colonial birds should be cognizant of these complications and plan their sampling procedures accordingly.

Addendum.—In 1986 Mono Lake gulls expanded their nesting locations on Negit Island and Paoha Island, which were reoccupied in 1985. The owls concentrated their activity near their roost on Negit Island, which lessened predation pressure elsewhere. On the Paoha Islets and Paoha Island sixteen adults (0.2% of the adult population) were killed between 17 April and 28 May. As in earlier years, most kills (14) were made on Coyote A and females suffered most (75%) of the mortality. Chick mortality was almost unrecorded. Three eagle kills of adult gulls (two males, one female) were made in areas of reduced visibility on Coyote A and Paoha Island. Interestingly, gulls nested successfully on Paoha Island, even though at least one coyote had access to the nesting area.

Antero Reservoir was drained and nesting was disrupted. Two adult gulls killed by eagles early in the season were the only evidence of avian predation.

ACKNOWLEDGMENTS

Jehl's studies were supported by the Los Angeles Department of Water and Power, under a cooperative agreement with the U.S. Fish and Wildlife Service, the National Geographic Society, and Sea World Research Institute. Chase's studies received financial support from the World Wildlife Museum of Monument, Colorado, and logistic support from the Denver Museum of Natural History. Permission to carry out research at Mono Lake was granted by the U.S. Forest Service through N. Upham. We acknowledge the cooperation of the San Diego State University Life Science Computer Facility for providing data management, analysis and reporting support. We are especially grateful to S. I. Bond for assistance in gathering and analyzing data and preparing the manuscript, and to J. L. Hand for bibliographic assistance. More than 50 volunteers cooperated in many ways over the years to make this study possible. M. Conover, K. Vermeer, W. Southern, and S. A. Mahoney reviewed the manuscript and made many helpful suggestions for its improvement.

LITERATURE CITED

BEHLE, W. H. 1958. The bird life of Great Salt Lake. University of Utah Press, Salt Lake City.

BUCKLEY, F. G., AND P. A. BUCKLEY. 1980. Habitat selection and marine birds. Pp. 69–112 *in* J. Burger, B. L. Olla, and H. E. Winn (eds.). Behavior of marine animals. Vol. 4: Marine Birds. Plenum Press, New York and London.

BURGER, J., AND M. GOCHFELD. 1984. Mobbing behavior in Common Terns: effects of colony size. Abstract. 102nd Ann. Meet. Amer. Ornith. Union.

CHARDINE, J. W., AND R. D. MORRIS. 1983. Nocturnal desertion patterns: influence on hatching synchrony in Ring-billed Gulls *Larus delawarensis*. Ibis 125:389–396.

EMLEN, J. T., D. E. MILLER, R. M. EVANS, AND D. H. THOMPSON. 1966. Predator-induced parental neglect in a Ring-billed Gull colony. Auk 83:677–679.

FINDHOLT, S. 1986. Status and distribution of California Gull nesting colonies in Wyoming. Great Basin Nat. 46:128–133.

JEHL, J. R., JR. 1987. Geographic variation and evolution in the California Gull (*Larus californicus*). Auk, in press.

JEHL, J. R., JR., AND S. A. MAHONEY. 1983. Possible sexual differences in foraging patterns in California Gulls and their implications for studies of feeding ecology. Colonial Waterbirds 6:218–220.

JEHL, J. R., JR., AND S. A. MAHONEY. MS. The role of thermal environment and predation in habitat choice in the California Gull.

KRUUK, H. 1964. Predators and anti-predator behavior of the Black-headed Gull (*Larus ridibundus*). Behav. Suppl. 11:1–129.

KRUUK, H. 1966. Defense against killers. Natural History 75:48–55.

LACK, D. 1954. The natural regulation of animal numbers. Clarendon Press, Oxford.

MARTIN, G. E. 1986. Sensory capacities and the nocturnal habit of owls (Strigiformes). Ibis 128:266–277.

MECH, D. L. 1970. The wolf. Natural History Press, New York.

METZGAR, L. H. 1967. An experimental comparison of Screech Owl predation on resident and transient White-footed Mice (*Peromyscus leucopus*). J. Mamm. 48:387–391.

MUELLER, H. C. 1974. Factors influencing prey selection in the American Kestrel, Auk 91:705–721.

NISBET, I. C. T. 1975. Selective effects of predation in a tern colony. Condor 77:221–226.

NISBET, I. C. T., AND M. J. WELTON. 1984. Seasonal variations in breeding success of Common Terns: consequences of predation. Condor 86:53–60.

PEARSON, O. P., AND A. K. PEARSON. 1947. Owl predation in Pennsylvania with notes on the small mammals of Delaware County. J. Mamm. 28:137–147.

PUGESEK, B. H., AND K. L. DIEM. 1983. A multivariate study of the relationship of parental age to reproductive success in California Gulls. Ecology 64: 829–839.

RUDEBECK, G. 1950–1951. The choice of prey and modes of hunting of predatory birds with special reference to their selective effect. Oikos 2:65–88; 3: 200–231.

SARGEANT, A. B., S. H. ALLEN, AND T. EBERHARDT. 1984. Fox predation on breeding ducks in midcontinent North America. Wildl. Mono. No. 89.

SOUTHERN, L. K., S. R. PATTON, AND W. E. SOUTHERN. 1982. Nocturnal predation on *Larus* gulls. Colonial Waterbirds 5:169–172.

SOUTHERN, W. E., S. R. PATTON, L. K. SOUTHERN, AND L. A. HANNERS. 1985. Effects of nine years of fox predation on two species of breeding gulls. Auk 102: 827–833.

VERMEER, K. 1970. Breeding biology of the California and Ring-billed gulls. Canad. Wildl. Serv. Rept. Series No. 12.

Studies in Avian Biology No. 10:102, 1987.

PROXIMATE MECHANISMS AFFECTING DIETARY SWITCHES IN BREEDING GULLS

CYNTHIA ANNETT[1]

ABSTRACT.—The diet of Western Gulls (*Larus occidentalis*) nesting on Alcatraz Island (San Francisco County) was followed for the breeding seasons of 1983, 1984, and 1985. Individual nests were surveyed two to four times per week throughout the breeding seasons, and data were collected on breeding biology, food remains, pellets, and regurgitations. Adults fed heavily on garbage early in the season, but switched to small fish (anchovy) when chicks hatched. This pattern was significant in all years (125–150 nests checked each year). The strong correlation of the timing of the switch with hatching of the eggs was supported by detailed data on the diet of individual pairs immediately before and after their eggs hatched. Experiments in which eggs were switched between nests in order to alter the timing of hatching also supported the hypothesis. Hatching alters many aspects of parental behavior, and appears to strongly influence prey choice behavior. Several hypotheses on the mechanism of prey switching as well as the evolutionary basis of prey choice by breeding gulls are discussed.

[1] Department of Zoology and Museum of Vertebrate Zoology, University of California, Berkeley, California 94720.

Studies in Avian Biology No. 10:103, 1987.

DIETS OF GLAUCOUS-WINGED GULLS: A COMPARISON OF METHODS FOR COLLECTING AND ANALYZING DATA

DAVID B. IRONS[1]

ABSTRACT.—Scientists collect data on the diets of gulls for many purposes. Depending on the objectives and the available resources, one or more of four main methods of determining diets may be best. All four have advantages and biases. My results showed that crop and/or gizzard contents provide some intact organisms, but crop contents are sometimes lost by regurgitation. Observations supply data for specific times and locations, but some prey cannot be identified. Regurgitated pellets furnish sizes of organisms and a large sample size, but are biased against soft-bodied prey. Food fed to chicks can be obtained easily but differs from adults' diets.

I analysed data from my studies using six methods: number, dry weight, volume, frequency of occurrence, an index of relative importance, and caloric value. All methods showed similar results but, depending on the type of prey eaten, results could vary greatly. The manner of collecting data, the study objectives, and the prey types should determine what method of analysis is used.

[1] United States Fish and Wildlife Service, 1011 E. Tudor Road, Anchorage, Alaska 99503.

Studies in Avian Biology No. 10:104, 1987.

PREDATORY BEHAVIOR OF YELLOW-FOOTED GULLS TOWARD HEERMANN'S GULL CHICKS AT DENSE AND SCATTERED NESTING SITES

ENRIQUETA VELARDE[1]

ABSTRACT.—Predation has been widely hypothesized as one of the factors having a major influence on group formation in birds. I studied the predatory behavior of Yellow-footed Gulls (*Larus livens*) toward Heermann's Gull (*L. heermanni*) chicks at dense and scattered nesting areas at Isla Rasa, Baja California. Gulls on dense areas had a significantly greater search time and number of predation attempts, obtained more prey but at a lower proportion than those on the more open areas. Although Heermann's Gull chicks cannot be a major food item for Yellow-footed Gulls, this predator may have an influence by selecting dense nesting in the Heermann's Gull. Topographic conditions may determine nesting density and set a limit to nest site availability for this species on islands of this area.

[1] Instituto de Biologia, Universidad Nacional Autonoma de Mexico, Apartado Postal 70-153, 04510 Mexico, D.F.

Studies in Avian Biology No. 10:105–118, 1987.

HABITAT AND NEST-SITE SELECTION OF MEW AND GLAUCOUS-WINGED GULLS IN COASTAL BRITISH COLUMBIA

KEES VERMEER[1] AND KEVIN DEVITO[2]

ABSTRACT.—The types of lakes, islands and nest sites Mew Gulls (*Larus canus*) selected on Vancouver Island, British Columbia, were examined. Mew Gulls nested on lakes that had significantly more islands and rivers, were larger, had longer shore lines, were closer to the sea, and were at lower altitudes than lakes not used by gulls. The nesting population of Mew Gulls on Vancouver Island was estimated from a highly significant regression equation containing these lake variables.

Mew Gulls significantly preferred small islands without trees, with a substrate of bare rock or low herbaceous cover. Nests were strongly associated with island crests and with a flat moss-covered substrate, and were significantly higher and farther from the water than randomly chosen matched points. Mew Gulls also nested frequently on tops of poles, pilings or tree stumps emerging from the water. On Vancouver Island, Mew Gulls nested mostly (80%) as solitary pairs.

Island and nest site selection by Mew Gulls was compared with that of Glaucous-winged Gulls (*Larus glaucescens*), a colonial nester at southeastern Vancouver Island. Ground-nesting Glaucous-winged Gulls selected islands that were significantly smaller, had a lower relief, were less forested and contained a higher percentage of bare rock or ground with low herbaceous cover than islands where gulls were absent. Nests of solitary pairs of Glaucous-winged Gulls, like Mew Gulls, were strongly associated with island crests, but unlike Mew Gulls, these had a grass substrate. Colonial Glaucous-winged Gulls showed no association with any island location, but were strongly associated with a bare rock substrate and, to a lesser extent, with grass. Matched points taken from both solitary and colonial nests occurred on significantly steeper slopes than nests and were also closer to shore.

Interspecific variation and plasticity in the nest site establishment of Mew and Glaucous-winged gulls are explored and their implications discussed.

The breeding range of Mew Gulls (*Larus canus*) in Eurasia extends as a broad band from northwestern Europe to Kamchatka in the eastern USSR, but is restricted to the northwestern portion of North America (Voous 1960). Atlases on breeding distribution of birds do not show the Mew Gull nesting in central and southern British Columbia (e.g., Voous 1960, Godfrey 1966), although it is a widespread, albeit dispersed breeder there. Campbell (1970) made the first attempt to compile information on the breeding status of Mew Gulls in British Columbia, but quantitative data were lacking at the time. Aspects of the nesting biology of Mew Gulls in British Columbia were described by Vermeer and Devito (1986). In this paper, we examine which lakes, islands, and sites Mew Gulls selected for nesting on Vancouver Island, British Columbia.

Island and nest site selection by Mew Gulls is compared with that of Glaucous-winged Gulls (*Larus glaucescens*) in the Gulf Islands at southeastern Vancouver Island. These species are the only coastal nesting larids in British Columbia. Glaucous-winged Gulls nest almost completely in the marine habitat in British Columbia (Drent and Guiguet 1961). The nesting biology of Glaucous-winged Gulls has been studied by Vermeer (1963), Ward (1973) and Hunt and Hunt (1976), but some aspects of its selection of islands and nest sites will be dealt with for the first time in this paper.

We also compare the nesting habitat of Mew Gulls on Vancouver Island with that of this species in northern Europe. The comparison will be subjective, because although the habitat of the species in Europe has been described (e.g., Bianki 1967, Haftorn 1971, Vaisanen and Jarvinen 1977, Götmark 1982), statistical treatment of nest-site selection similar to our method is lacking.

METHODS

Surveys of nesting Mew Gulls on Vancouver Island were conducted during the last week of May and first half of June in 1984 and 1985 on 58 lakes accessible by road (Table 1, Fig. 1). Lakes were chosen from those present on a Canadian Hydrographic Service (CHS) chart (Chart No. 4C 3001, scale 1:525,000). The entire shoreline of each lake was explored by boat and the position of each nest was mapped. The number of islands present was noted in the field, while measurements of lake area, perimeter, altitude, and distance to sea were made from 1:50,000 and 1:80,000 scale maps. Mean lake depths and supporting morphometric data for 40 lakes were obtained from lake inventory files of the Fisheries Branch, Ministry of Environment, Victoria, British Columbia. The number of rivers entering or leaving a lake was determined by counting the blue lines indicating the water courses on the CHS chart. The extent of human disturbance at a lake was determined by a method similar to that described for loons

[1] Canadian Wildlife Service, P.O. Box 340, Delta, British Columbia V4K 3Y3, Canada.

[2] Watershed Ecosystem Program, Trent University, Peterborough, Ontario K9J 7B8, Canada.

TABLE 1

LAKES ON VANCOUVER ISLAND SURVEYED FOR NESTING MEW GULLS IN MAY AND JUNE OF 1984 AND 1985

Name of lake	Number of nests	Name of lake	Number of nests
1. Georgie Lake	0	30. Mohun Lake	2
2. Nahwitti Lake	0	31. Morton Lake	0
3. Kains Lake	3	32. Amor Lake	1
4. Quatse Lake	0	33. Gray Lake	0
5. Alice Lake	4	34. Boot Lake	0
6. Victoria Lake	1	35. Goslin Lake	0
7. Benson Lake	0	36. John Hart Lake	0
8. Maynard Lake	0	37. Campbell Lake	2
9. Iron Lake	0	38. Beavertail Lake	0
10. Lac Truite	0	39. Gooseneck Lake	1
11. Three Isle Lake	4	40. Middle Quinsam Lake	0
12. Keogh Lake	0	41. Wokas Lake	0
13. Nimpkish Lake	4	42. Upper Quinsam Lake	0
14. Atluck Lake	0	43. Upper Campbell Lake	0
15. Wolfe Lake	0	44. Buttle Lake	0
16. Mukwilla Lake	0	45. Wolf Lake	0
17. Woss Lake	0	46. Comox Lake	1
18. Klaklakama Lake 1	0	47. Elsie Lake	0
19. Klaklakama Lake 2	0	48. Great Central Lake	2
20. Schoen Lake	0	49. Sproat Lake	3
21. Vernon Lake	0	50. Horne Lake	1
22. Muchalat Lake	0	51. Cameron Lake	0
23. McCreight Lake	0	52. Muriel Lake	3
24. Pye Lake	0	53. Kennedy Lake	24 (27)*
25. Lower Stella Lake	0	54. Maggie Lake	3
26. Stella Lake	1	55. Nanaimo Lake 1	0
27. Roberts Lake	1	56. Nanaimo Lake 2	0
28. Mud Lake	0	57. Cowichan Lake	11
29. Amor Lake	1	58. Nitinat Lake	0

* 24 nests in 1984 and 27 nests in 1985.

by Vermeer (1973). Ten disturbance units were given for each patrolled campsite and resort, five units for each wilderness campsite (camp area or a spot for 1 or 2 campers), and one unit for each house and summer cottage bordering a lake. For each nesting island, the surface area and the distance to the mainland shore and nearest island were determined from maps or actual site measurements. Presence of trees (>1.5 m tall) and percent bare ground (low herbaceous cover and/or bare rock) were noted. At Kennedy Lake, measurements were taken at each island for later comparison between islands with and without nests. The dominant vegetation, distance to shore, nest height above water, distance to nearest tree, and microslope were determined for nests and for matched points 2 m away from nests. Microslope was determined from the average maximum slope over a 30 cm distance centered at the nest or matched point. Matched points were selected randomly from one to 10 compass points, similar to the method described by Burger and Gochfeld (1981). The distance of 2 m was chosen arbitrarily, but probably fell within the nesting territory as the distance to the nearest neighboring nest, where more than one pair of gulls nested on an island, ranged from 3.5 to 15 m.

Surveys on nesting Glaucous-winged Gulls were conducted in the Gulf Islands, from 49°09′N (including the Flat Top Islands) to 48°33′N (D'Arcy Island) during June and July 1985. All islands were visited by boat, the number of nesting gulls was counted at each nesting island and the number and position of each colony or solitary nest site were mapped. For each island the percent bare ground (low herbaceous cover and/or bare rock) and extent of forestation (>50 percent = forested island) was noted. The surface area, distance to mainland (Vancouver Island) and maximum height of each island were determined from the CHS chart No. 3310 (sheets 1 to 4) or at the site.

Observations similar to those made on Mew Gulls were made on the nest sites of Glaucous-winged Gulls. Approximately 5% of the ground nests (up to a maximum of 10 nests) were selected randomly from each colony. Matched points were selected 2 m from solitary nests and the selected colonial nesting pairs. Since only a small number of solitary Glaucous-winged Gull pairs were located on islands (there were more on beacons and cliffs), one island with three and another with two pairs were included in the single pair category. Those two islands were large and the pairs nested far apart in a solitary manner.

For both lake and island parameters, an overall or multivariate test was conducted to test the hypothesis of equal population mean vectors for lakes and islands with and without nests (see Hummel and Sligo 1971). The Wilks Lambda statistic (λ) was used. Following a

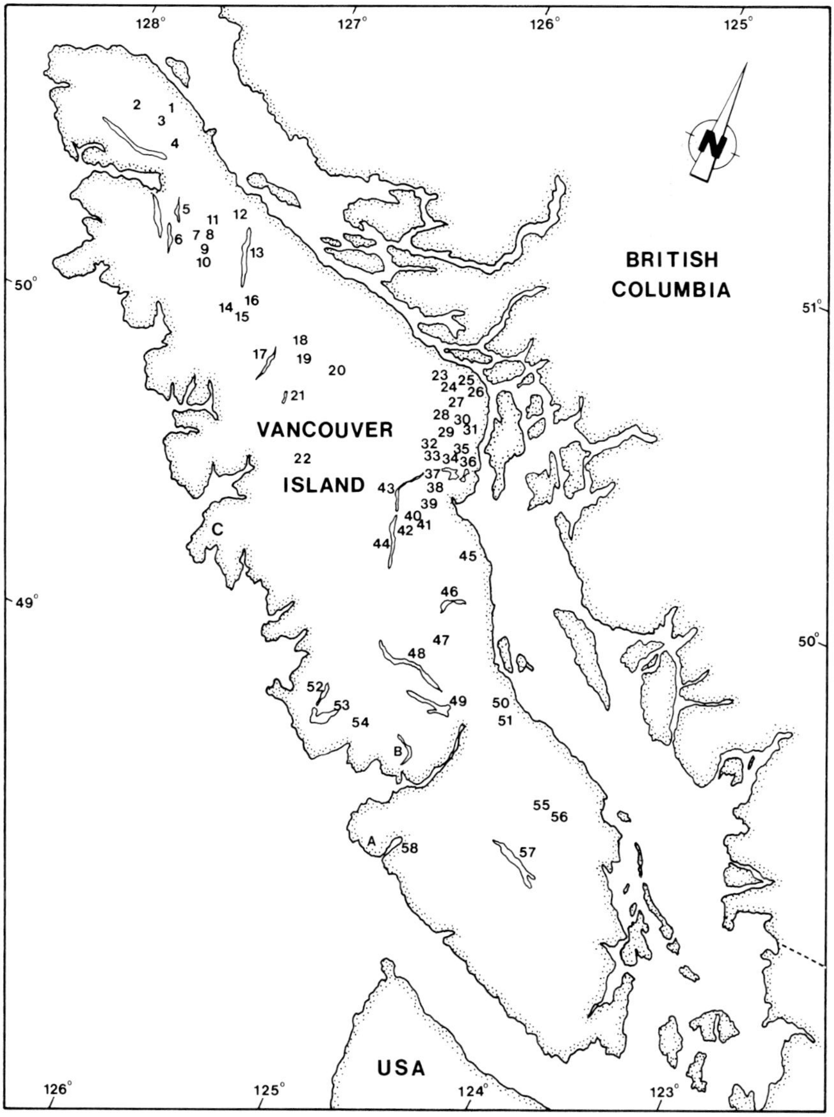

FIGURE 1. Lakes surveyed and those with nesting Mew Gulls on Vancouver Island, 1984 and 1985. A = Tsusiat lake, B = Henderson Lake, C = Hesquiat Lake.

significant overall difference, a univariate ANOVA was run on each variable separately. At Kennedy Lake, classification of islands with and without nests was determined by Discriminant Analysis. Relationships between the number of nests per lake and other lake variables were determined from regression analyses. Data analyses were performed using the SPSSx statistical package at the University of Victoria.

Association of nest position and absence or presence of vegetation for nests of Mew and Glaucous-winged gulls and their matched points was determined by Chi-square analyses. Distance measurements to nests and matched points were analyzed using the normal approximation of the Sign Test (see Zar 1974).

RESULTS

Nest Dispersion and Population of Mew Gulls

Seventy-six nests with eggs or small chicks were found on islands (38), on tops of poles and tree stumps in the water (29), and in trees (9) on 20 of the 58 lakes (Table 1, Fig. 1). All nests on poles, tree stumps, and in trees were single nests. Of the 38 nests on islands, 23 were solitary, one island had 4, three islands 3, and one island 2 nests. The Mew Gull, with 80 percent of the nests

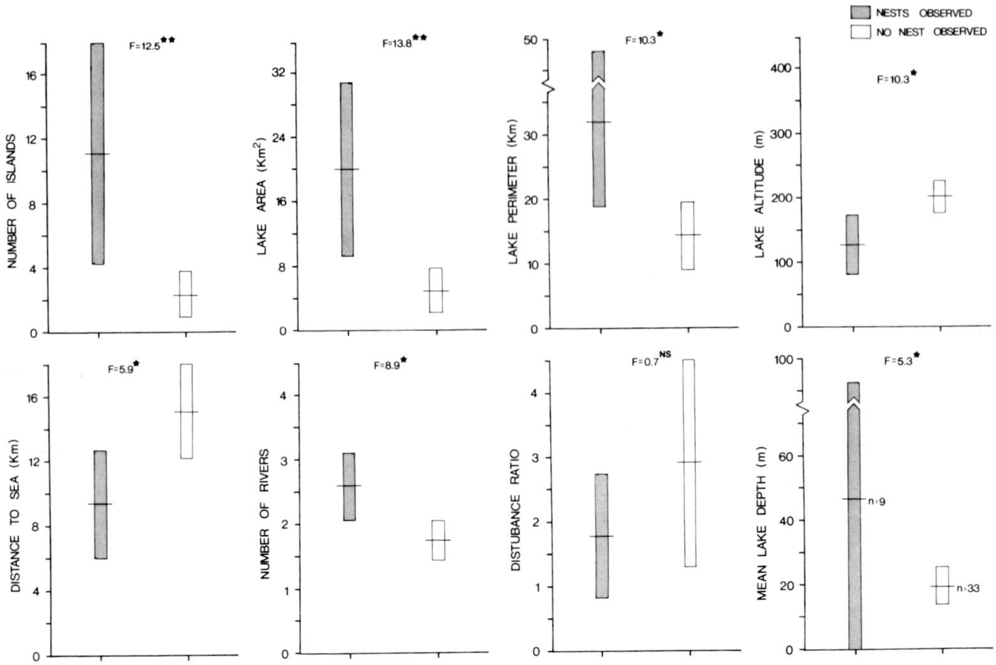

Figure 2. Means (horizontal lines) and 95 percent confidence intervals (rectangles) of variables of lakes with (n = 19) and without (n = 39) nesting Mew Gulls on Vancouver Island, 1984 and 1985. Overall significance with multivariate test: Wilks Lambda = 0.658, F = 3.715, $P = 0.003$. Mean lake depth not included in multivariate test. Values of univariate F test (df = 1,56) are shown on Figure. $*P < 0.05$, $**P < 0.001$.

found being single, is therefore predominantly a solitary nester on Vancouver Island.

To estimate the total nesting population of Mew Gulls on Vancouver Island, the numbers of observed Mew Gull nests (NN) on lakes were regressed on the following lake variables: number of islands present (NI), lake area (LA), lake perimeter (LP), lake altitude (ALT), distance of a lake to sea (DS), and the number of rivers associated with a lake (NR). The estimated regression equation was:

$$NN = 1.07 + 0.127\ NI + 0.276\ LA - 0.155\ LP + 0.001\ ALT - 0.039\ DS + 0.125\ NR.$$
$$(F_{6,51} = 68.79, \quad R^2 = 0.890, \quad \hat{R}^2 = 0.877, \quad P < 0.0001)$$

Of a total of 123 lakes shown on the CHS chart, 58 were surveyed, and 65 lakes were not surveyed for nesting Mew Gulls in 1984 and 1985. Of the latter, Tsusiat, Henderson and Hesquiat lakes were known, from previous surveys, to have together 7 pairs of nesting gulls (B.C. Prov. Museum records). Using the above regression, it was predicted that 44 pairs of gulls nested on the 62 lakes not surveyed and lacked nesting records. The overall nesting population of nesting Mew Gulls on Vancouver Island lakes was estimated at 127 pairs for 1984 and 1985.

Habitat Selection by Mew Gulls

A multivariate test for seven variables showed an overall significant difference between lakes with and without nesting Mew Gulls (Fig. 2). Univariate tests for each separate variable in-

TABLE 2

Discriminant Analysis between Lakes with and without Mew Gulls on Vancouver Island, Using Number of Islands, Lake Area, Lake Perimeter, Altitude, Distance to Sea and Number of Rivers

Observation		Prediction	
Lake category	Number of lakes	Number of lakes with nests (%)	Number of lakes without nests (%)
Lakes with nests	19	15 (78.9)*	4 (2.1)
Lakes without nests	39	7 (17.9)	32 (82.1)*

Discriminant function = 0.492 NI − 0.022 LA + 0.47 LP + 0.002 ALT − 0.538 DS + 0.304 NR. Wilks Lambda = 0.661, $\chi^2 = 21.93$, 6 df, $P = 0.0012$.

* Lakes correctly classified at 47/58 = 81%.

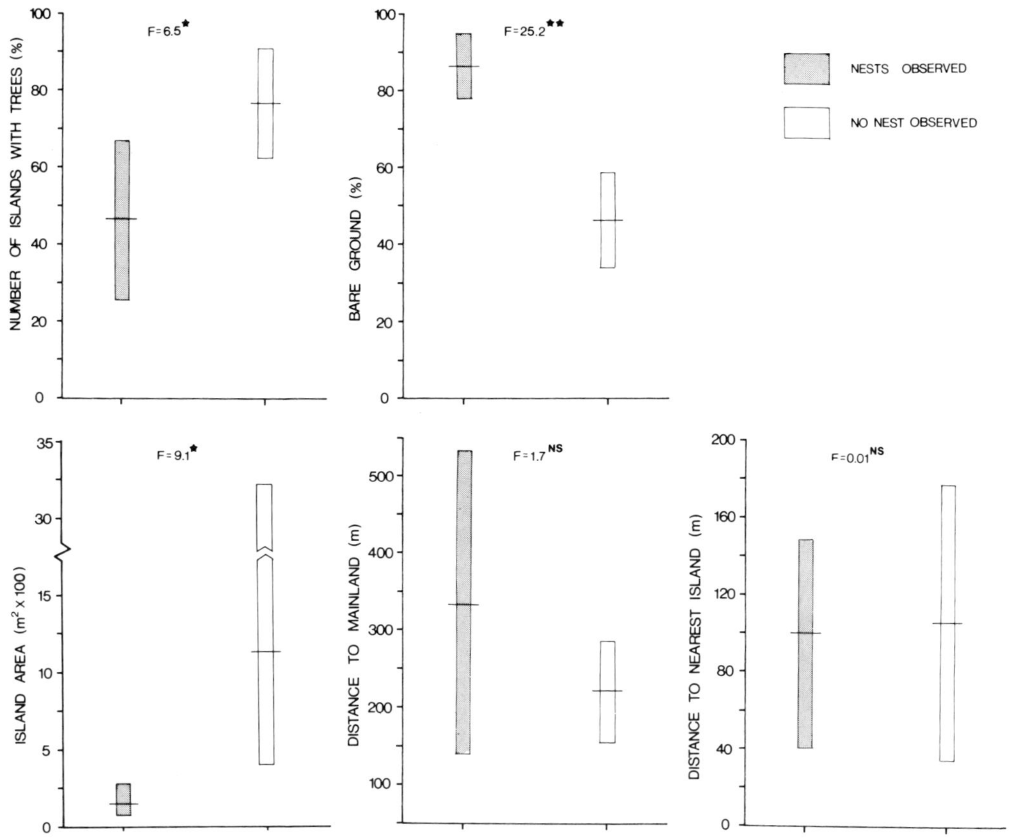

FIGURE 3. Means and 95 percent confidence intervals of variables of islands with (n = 26) and without (n = 38) nesting Mew Gulls on Kennedy Lake, 1984 and 1985. Overall significance with multivariate test: Wilks Lambda = 0.690, F = 5.206, P = 0.01. Univariate F test (df = 1,62).

dicated that Mew Gulls nested on lakes which had significantly more islands and rivers, were larger, and had a longer shoreline than lakes where gulls were absent (Fig. 2). Gulls also nested significantly more often on low altitude lakes and close to the sea. Since lakes at low altitude on Vancouver Island are close to the sea, those two variables are interdependent (r = 0.672). Of the seven variables tested separately, only estimated extent of human disturbance did not appear to be a significant limiting factor to nesting Mew Gulls (Fig. 2). Because of the smaller sample size than for the previously mentioned seven variables, mean lake depth was not included in the multivariate test. Of the lakes analyzed, those with nesting gulls appeared to be deeper (Fig. 2). The discriminant function of the six significant lake variables correctly classified 81 percent of the lakes surveyed (Table 2).

Islands used by nesting Mew Gulls were smaller, had a greater percentage of bare ground, and usually lacked trees (Fig. 3). Nesting by Mew Gulls at 75 percent of the islands could be predicted on the basis of the latter two variables alone (Table 3). There was no significant difference between nesting and non-nesting islands with respect to the nearest distance to the mainland or the nearest neighboring island.

Comparison of island nest sites with their matched points indicates that nests were strongly associated with island crests which have a moss substrate whereas matched points were associated with island slopes and peripheries on moss or bare rock (Table 4). Nests were situated on nearly flat surfaces (2°), which usually contained a small depression to secure them, while their matched points occurred on significantly steeper slopes (30°) (Fig. 4). Nests also were significantly farther and higher from the water than their matched points. Gulls nested significantly farther apart when more than one pair was present than if nests had been randomly distributed (Fig. 4).

Poles, pilings, or tree stumps used by nesting Mew Gulls frequently were joined and about 2

TABLE 3

DISCRIMINANT ANALYSIS BETWEEN ISLANDS WITH AND WITHOUT MEW GULLS ON KENNEDY LAKE, 1984 AND 1985, USING PRESENCE OF TREES AND PERCENT BARE GROUND

Observation		Prediction	
Island category	Number of islands	Number of nesting islands (%)	Number of non-nesting islands (%)
Nesting islands	26	22 (84.6)*	4 (15.4)
Non-nesting islands	38	12 (31.6)	26 (68.4)*

Discriminant function = 0.178 tree + 1.0999 bare ground. Overall significance tested, Wilks Lambda = 0.707, $\chi^2 = 21.11$, 2 df, $P < 0.005$.

* Islands correctly classifed at 48/64 = 75%.

TABLE 4

NEST POSITION AND ABSENCE OR PRESENCE OF VEGETATION FOR 42 NEST SITES AND THEIR MATCHED POINTS OF MEW GULLS ON VANCOUVER ISLAND, 1984 AND 1985

Variables measured	Number of nest sites (percentage in parentheses)	Number of matched points (percentage in parentheses)
Nest position		
On crest of island	39 (92.9)	4 (9.5)
On flat top of island	1 (2.4)	0 (0.0)
On slope of island	2 (4.7)	27 (64.3)
On island periphery	0 (0.0)	11 (26.2)
Chi-square values[1]	196.3**	40.5**
Chi-square values	62.0**	
Absence/presence of dominant vegetation		
Bare rock	6 (14.6)	21 (51.2)
Moss	27 (65.9)	13 (31.7)
Grass	5 (12.2)	4 (9.8)
Shrub	3 (7.3)	3 (7.3)
Chi-square values	36.95**	20.95**
Chi-square values	16.34**	

[1] Shown are Chi-square values with 3 degrees of freedom.
** $P < 0.001$.

to 2.5 m above water (Table 5). Trees used for nesting were moderate-sized coniferous trees close to shore. The nests were on limbs not far from the trunk or on broken tops at an average height of 9 m above the water surface (Table 5).

NEST DISPERSION AND POPULATION OF GLAUCOUS-WINGED GULLS

Glaucous-winged Gulls nested primarily in colonies and on the ground. A total of 3393 nesting pairs was found at 46 sites (Fig. 5). At these sites, 3269 pairs nested on the ground at islands, 119 pairs on cliffs (some in cliff cavities), 4 pairs on beacons, and 1 pair on a piling. Only 0.5 percent were observed to be solitary nesters.

HABITAT SELECTION BY GLAUCOUS-WINGED GULLS

Glaucous-winged Gulls nested on islands of various sizes but a majority of the population nested on islands 2 to 10 ha in size (Fig. 6). Ground nesting occurred on small to moderate sized islands but cliff nests predominated on large islands (Fig. 6).

A multivariate test for six island variables showed significant difference between islands with and without Glaucous-winged Gulls (Table 6). Univariate tests for each separate variable indicated that islands with gulls were significantly less forested and had more bare ground and cliffs than those without gulls (Table 6). No significant differences were observed in island area, height and distance of island to nearest mainland. That there was no significant difference in island size between nesting and non-nesting islands can be explained in that gulls nesting on cliffs generally nested on much larger islands, and were not as restricted by island size as ground-nesting gulls (Fig. 6). Since the vast majority of Glaucous-winged Gulls nested on the ground, a separate analysis of variance was carried out for islands without cliffs. Univariate tests for each separate variable indicated that ground-nesting Glaucous-winged Gulls selected islands which were significantly smaller, had a lower relief, were not forested and contained a higher percentage of bare ground than islands where gulls were absent (Fig. 7). Of the five variables tested separately, only distance to the mainland did not differ significantly between nesting and non-nesting islands.

Comparison of island nest sites with their matched points indicated that the nests of solitary Glaucous-winged Gulls, like Mew Gulls, were strongly associated with island crests, while their matched points showed no association with any position (Table 7). Nests of solitary pairs were strongly associated with grass as opposed to their matched points which were always on bare rock. Nests of colonial pairs showed no association with any position variable, but were strongly associated with bare rock and, to a lesser extent with grass, while matched points mostly occurred on bare rock. Matched points of both solitary and colonial nests occurred on significantly

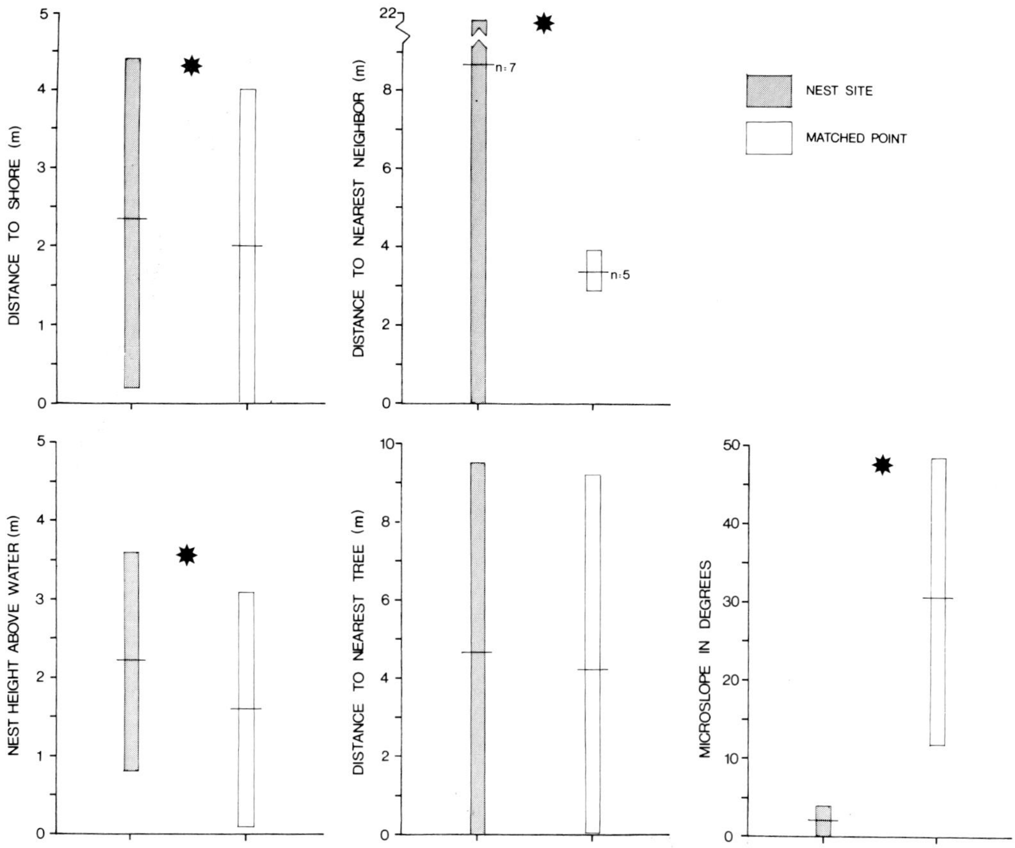

FIGURE 4. Means and 95 percent confidence intervals of variables of 42 island nest sites and their 42 matched points of Mew Gulls on Vancouver Island, 1984 and 1985. An asterisk indicates significant difference ($P < 0.05$) by normal approximation of the Sign Test.

steeper slopes and were closer to shore than nest sites (Fig. 8). No difference was observed between nests and matched points for distance to nearest neighboring nest and height above water.

DISCUSSION

Nest Dispersion and Population of Mew Gulls

Nest dispersion

One of the factors determining nest dispersion in Mew Gulls is the physical aspect of the nesting habitat. Other factors which may affect nest dispersion or extent of coloniality in gulls are predation, social factors, and food availability (e.g., Burger and Shisler 1980, Burger 1981, Andersson et al. 1981, Götmark 1982, and Götmark and Andersson 1984), but their discussion falls outside the scope of this paper. If a nesting substrate consists of a small rock, pole or tree stump, there is no room for colony establishment. On Vancouver Island most of the lake islands with nesting Mew Gulls were very small and devoid of trees. Moderate-sized islands usually were densely forested. The relative scarcity of moderate-sized islands without trees may have been one of the factors determining the predominant solitary nesting pattern of Mew Gulls. Götmark and Andersson (1984) reported that Common (=Mew) Gulls, on lakes with densely wooded islands in southern Sweden, avoided breeding among trees and were forced to nest solitarily on scattered boulders and small skerries.

Population

We estimated the total population of Mew Gulls on Vancouver Island at 127 pairs. Campbell (1970), who reviewed the nesting status of Mew Gulls in British Columbia 16 years ago, reported only three known nesting lakes on Vancouver Island: Cowichan, Kennedy, and Sproat lakes. At present there are 23 known nesting lakes

TABLE 5

NESTING PARAMETERS FOR MEW GULLS ON TOPS OF POLES, PILINGS AND TREE STUMPS EMERGING FROM THE WATER AND IN TREES ALONG LAKE SHORES ON VANCOUVER ISLAND, 1984 AND 1985

Location of nest, parameters	Number of nests	Mean and standard deviation
Nests on poles, pilings, tree stumps		
Number of poles/stumps in a group	24	1.96 ± 1.81
Area of nesting platform (m^2)	23	1.07 ± 1.14
Nest height above water (m)	24	2.39 ± 1.71
Distance to mainland (m)	24	62.6 ± 96.8
Distance to nearest pole or island (m)	24	28.5 ± 42.8
Tree nests		
Distance of tree to shore (m)	9	3.07 ± 2.89
Nest height above water (m)	9	9.03 ± 4.96
Tree diameter (m)	5	0.38 ± 0.17
Tree height (m)	8	9.13 ± 5.52
Distance from tree trunk (m)	9	0.89 ± 0.80
Distance to nearest tree (m)	7	2.46 ± 1.40

(Fig. 1, Table 1). The increase in lakes used by Mew Gulls over the past 16 years may not represent a population increase on Vancouver Island, but a better knowledge of nesting localities. Many nests found during this survey were on lakes not investigated previously.

To detect if the population is changing, we suggest that Mew Gulls on Vancouver Island be surveyed every 10 to 15 years. The nesting chronology of the species shows that the best time for conducting surveys is the first half of June, when most clutches have been laid, but have not yet hatched (Vermeer and Devito, 1986). Another practical reason for the suggested surveys is that lakes and rivers in British Columbia, including Vancouver Island, are being increasingly dammed for hydro-electricity and reservoirs. It appears that regulated lakes have fewer nesting Mew Gulls than natural ones (Vermeer and Devito, unpubl. data). Weselowski et al. (1984) also found that the damming of the Vistula River in Poland for reservoir purposes resulted in a withdrawal of terns and waders as well as a decline in numbers of gulls in the heavily changed section of the river.

HABITAT SELECTION BY MEW GULLS

The lakes most frequently selected by Mew Gulls were large, deep, and had many islands and several rivers. The large number of islands may have been the important feature. Larger lakes generally have more islands and, therefore, have more potential nest sites. Poles, pilings and tree stumps, which generally are most numerous on large lakes (due to logging activity), are also used as nest sites. We did not investigate the number of poles, pilings and tree stumps in a lake in relation to nest site requirements. Prior to our study Mew Gulls had not been recorded using those sites on Vancouver Island.

Many potential nesting islands, poles, and stumps were unused, suggesting that there are other requirements for nesting. In a study at Kennedy Lake, adult Mew Gulls fed mostly on crayfish in the lake and associated rivers (Vermeer and Devito 1986). This may explain the apparent importance of rivers, although the number of rivers is usually a function of lake size. Adults brought their young fish which were caught in the nearby marine intertidal zone. Mew Gulls rely, therefore, only partially on nesting lakes and their rivers for food, and the remainder of their diet comes from marine habitats. This may explain why many Mew Gulls nested close to the sea.

Mew Gulls nested predominantly on the crests of small treeless islands that were largely moss-covered and bare rock. Small bare islands may be less frequently patrolled by mammalian predators such as mink and river otter than the larger forested islands, and nesting on a crest may allow the Mew Gull to see any approaching predator from a distance.

HABITAT SELECTION BY GLAUCOUS-WINGED GULLS

Glaucous-winged Gulls in the Gulf Islands nested predominantly on the ground of small to moderate-sized islands. Only a small fraction of the population (4%) nested on cliffs, and this usually occurred on larger islands. Cliff nesting occurred on larger islands perhaps because (1) most cliffs occurred there, and (2) cliffs are inaccessible, whether on small or large islands, to

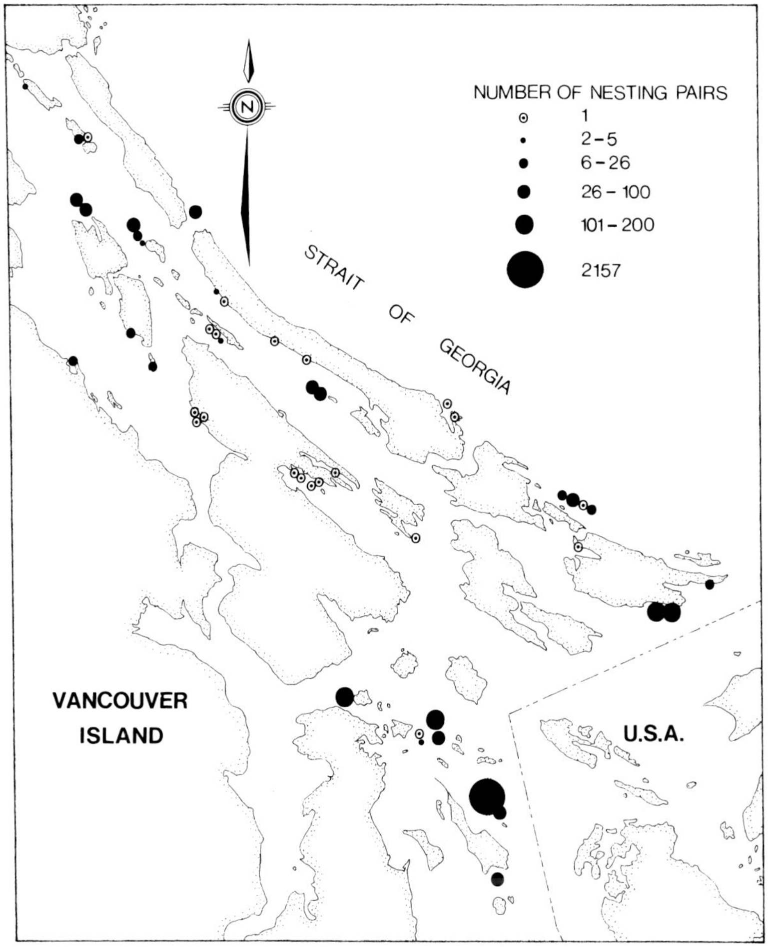

FIGURE 5. Nest dispersion of Glaucous-winged Gulls in the Gulf Islands, SE Vancouver Island, 1985.

mammalian predators such as the river otter (*Lutra canadensis*) (Foottit and Butler 1977, Verbeek and Morgan 1978). It is not surprising that ground-nesting gulls used relatively small islands with few trees. Most of the large forested islands had villages or residences with cats and dogs, and undoubtedly established populations of wild predators, which could prey upon eggs, chicks and possibly adults. Gulls also avoided small and moderate-sized islands that were forested, possibly because of difficulties they had in negotiating departure from and landing at nest sites among trees.

Solitary pairs of Glaucous-winged Gulls preferred island sites near crests for nesting. Colonial pairs, on the other hand, did not show a preference for any nest-site location, perhaps because nests in large and often saturated colonies are uniformly distributed. Intraspecific aggression at these latter locations may play a decisive role in the spacing of pairs and consequent site establishment (Vermeer 1970). Elevation in a colony is not as essential as it is for solitary nesters, as the alarm calls of gulls nesting on crests will warn those in more concealed conditions of the approach of predators. On Mandarte Island, a long-established and the largest colony in the Gulf Islands (Fig. 5), many gulls nest at the end of a tunnel underneath shrubs.

Interspecific Variation in Nest-site Selection between Mew and Glaucous-winged Gulls

Mew and Glaucous-winged gulls in British Columbia nest in very different macrohabitats, the former in fresh and the latter in marine waters. In Europe, however, Mew Gulls nest commonly in marine waters in association with larger Her-

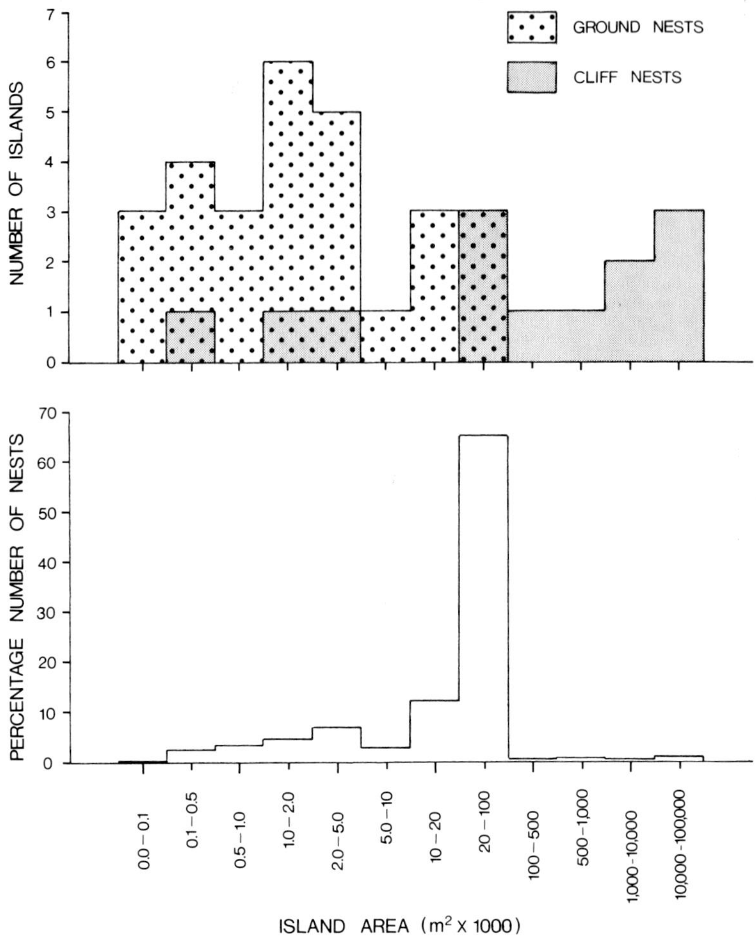

FIGURE 6. Comparison of numbers of nests and island size for ground and cliff-nesting Glaucous-winged Gulls, Gulf Islands, 1985.

ring Gulls (*Larus argentatus*) as well as Greater Black-backed Gulls (*Larus marinus*), although the different species have distinct habitat preferences (Götmark 1982, Hanssen 1984, see also Bianki 1967). The reason Mew Gulls are restricted to the fresh water habitat in British Columbia and not in Europe is unknown. Perhaps the Mew Gull is gradually expanding into a new habitat

TABLE 6

ANALYSIS OF VARIANCE OF MEANS OF VARIABLES OF ISLANDS WITH (N = 41) AND ISLANDS WITHOUT (N = 160) NESTING GLAUCOUS-WINGED GULLS IN THE GULF ISLANDS, 1985

Univariate F test (df = 1,199)	Mean square error	F statistic	Signif. of F
Variables			
Natural log of island area (m²)	20.82	2.22	0.138
Island height (m)	275.64	0.06	0.807
Natural log of distance to nearest mainland (m)	9.69×10^6	0.24	0.620
Forest present	3.58	15.26	<0.001
Bare ground (%)	3.46×10^4	24.46	<0.001
Cliffs present	2.64	24.52	<0.001
Overall significance with the multivariate test Wilks Lambda = 0.712, F = 13.066, $P < 0.001$			

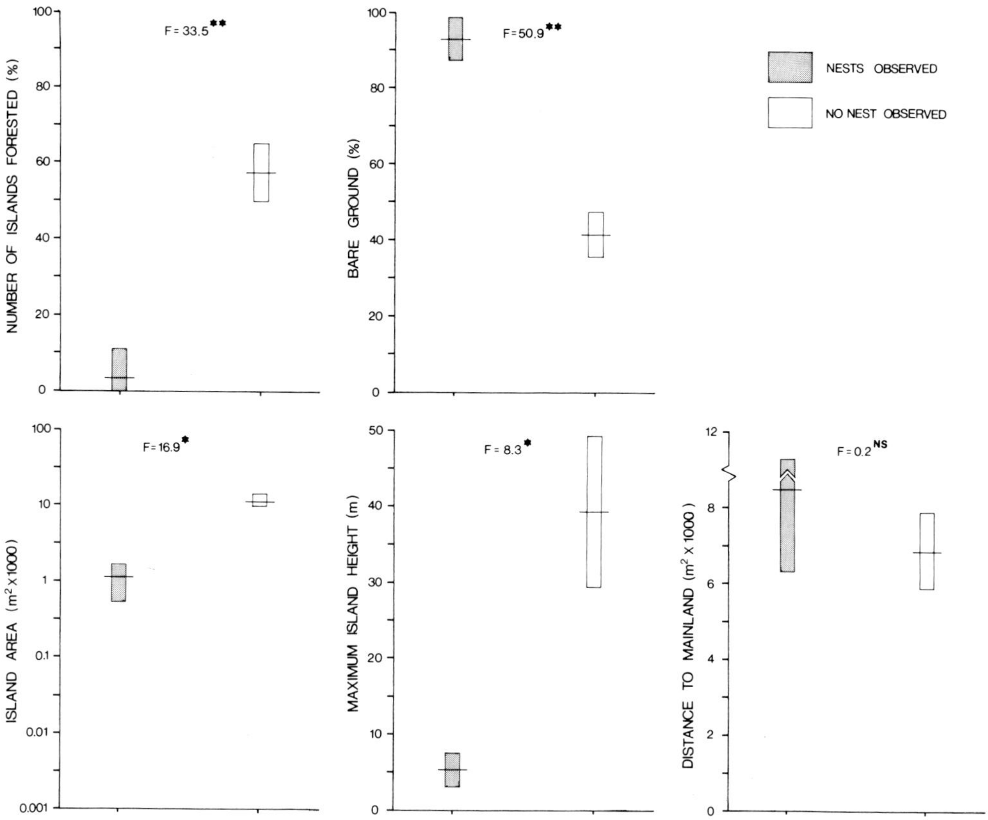

FIGURE 7. Means and 95 percent confidence intervals of variables of islands with (n = 29) and without (n = 160) ground-nesting Glaucous-winged Gulls in the Gulf Islands, 1985. Overall significance with multivariate test: Wilks Lambda = 0.761, F = 9.541, $P < 0.001$. Univariate F test (df = 1,87).

in British Columbia, because the niche in the marine environment is already filled by Glaucous-winged Gulls.

In some instances, Mew and Glaucous-winged gulls nest in different microhabitats (e.g., cliffs versus trees), but selection of island sites by solitary pairs is remarkably similar. Solitary pairs of both species nested on the crests of small islands with low herbaceous cover of grass or moss. Grass usually grows on the elevated interior of small islands used by Glaucous-winged Gulls, while most of the perimeter is bare rock subject to wave action during high tides and storms. Mew Gulls in Scandinavia, like Glaucous-winged Gulls in British Columbia, nest on grass substrates on low and small to moderate-sized islands in the marine environment (Vaisanen and Jarvinen 1977, Hanssen 1984). In contrast to the marine habitat, Mew Gulls nesting on rocky islands, poles, tree stumps, and pilings in Vancouver Island lakes, are usually safe from wave action and tides.

Plasticity in Habitat Selection of Mew and Glaucous-winged Gulls

Mew Gulls

Mew Gulls nest mostly on small, fresh water islets, on poles and tree stumps on large lakes in British Columbia (this study); on marshes and small islands in ponds, lakes and rivers in Alaska (Bent 1921, Strang 1974, Burger and Gochfeld pers. comm.); and on marine islands in the Gulf of Alaska (Hatch et al. 1978). In Scandinavia, they nest in bogs, on islands in ponds and lakes (Ytreberg 1956, Haftorn 1971, Götmark and Andersson 1984) as well as on islands in the sea (Götmark 1982, Hanssen 1984), and in trees and on roofs of houses (Haftorn 1971). In the interior of Poland, Mew Gulls nest extensively on river islands (Weselowski et al. 1984). On islands in the White Sea of the western USSR, Mew Gulls prefer to nest on maritime meadows and in areas with crowberries (Bianki 1967). In Scandinavia, they also nest in mountain heath (*Empetrum ni-*

TABLE 7

NEST POSITION AND ABSENCE OR PRESENCE OF VEGETATION FOR NEST SITES AND THEIR MATCHED POINTS OF 12 SOLITARY AND 42(44) COLONIAL PAIRS OF GLAUCOUS-WINGED GULLS IN THE GULF ISLANDS, SE VANCOUVER ISLAND, 1985

Variables measured	Solitary nesters		Colonial nesters	
	Number of nest sites	Number of matched points	Number of nest sites	Number of matched points
	(percentage in parentheses)			
Nest position				
On crest of island	9 (75.0)	3 (35.0)	12 (28.6)	8 (19.0)
On flat top of island	1 (8.3)	1 (8.3)	8 (19.0)	10 (23.8)
On slope of island	2 (16.7)	5 (41.7)	10 (23.8)	15 (35.7)
On island periphery	0 (0.0)	3 (35.0)	12 (28.6)	9 (21.4)
Chi-square values[1]	0.67*	2.67	1.05	2.76
Chi-square values	5.79		2.45	
Absence/presence of dominant vegetation				
Bare rock	3 (25.0)	12 (100.0)	24 (54.5)	38 (86.4)
Moss	0 (0.0)	0 (0.0)	0 (0.0)	0 (0.0)
Grass	9 (75.0)	0 (0.0)	15 (34.1)	4 (9.1)
Shrub	0 (0.0)	0 (0.0)	5 (11.4)	2 (4.5)
Chi-square values[1]	18.0**	36.0**	31.3**	89.9**
Chi-square values	14.4**		10.4**	

[1] Three degrees of freedom.
* $P < 0.05$, ** $P < 0.005$.

grum) in association with Arctic Skuas (*Stercorarius parasiticus*) (Götmark and Andersson 1980). Nesting in trees is common in British Columbia (this study) and in certain regions in Norway (Haftorn 1971), and apparently occasional in Alaska (Bent 1921, Dick et al. 1976), Finland (Silvola 1957) and the northeastern USSR (Bianki 1967). From the above observations it is evident that the nest habitat of Mew Gulls varies tremendously over its breeding range. The different nesting patterns of Mew Gulls may be the result of inherent behavioural differences among populations or may reflect species plasticity. At Stavanger in northern Norway, Mew Gulls nesting in mountain heath in association with Arctic Skuas had significantly smaller nests and less nest material than at a nearby colony on a lake island (Götmark and Andersson 1980). The gulls nesting in mountain heath, of which crowberry is an important component, apparently nest in similar habitat as those at Kandalaksha Bay, USSR. Bianki (1967) observed that Mew Gulls at Kandalaksha Bay nested primarily in maritime meadows with crowberry patches. In this habitat Mew Gulls commonly built on driftwood logs, hillocks, and small rocks, but when on dry level ground the nest site was often simply a hole with a meager lining of grass stems. Bianki's (1967) observation of the two nest types within a crowberry meadow suggests a high degree of plasticity within the same nesting population.

Glaucous-winged Gulls

Glaucous-winged Gulls, unlike Mew Gulls, usually are restricted to the marine habitat in the North Pacific Ocean (Drent and Guiguet 1961, Sowls et al. 1978). One pair of Glaucous-winged Gulls was observed nesting on top of a piling on Cowichan Lake during our survey of Mew Gulls. In Washington and Oregon, a small number of Glaucous-winged Gulls nested in colonies of California (*Larus californicus*) and Ring-billed gulls (*L. delawarensis*) about 300–500 km from the mouth of the Columbia River (Conover 1984). Inland-nesting of Glaucous-winged Gulls appears to be a recent phenomenon. Hybridization between Glaucous-winged and Herring gulls on inland lakes, however, has been observed in Alaska (Williamson and Peyton 1963) and in British Columbia (Merilees 1974).

Within the marine habitat, Glaucous-winged Gulls nest on islands, jetties, cliffs, beacons, pilings, bridge abutments, barges, derricks, log booms and occasionally in cliff cavities and large trees (Drent and Guiguet 1961, Campbell 1975, this study) as well as on roofs of buildings in coastal cities (Oldaker 1963, Eddy 1982, authors' unpubl. observations). The occupation of roofs by Glaucous-winged Gulls has increased during the last three decades and many hundreds of gulls presently nest on roofs in Seattle and Vancouver (Eddy 1982, Vermeer unpubl. observations). The above observations indicate that the Glaucous-

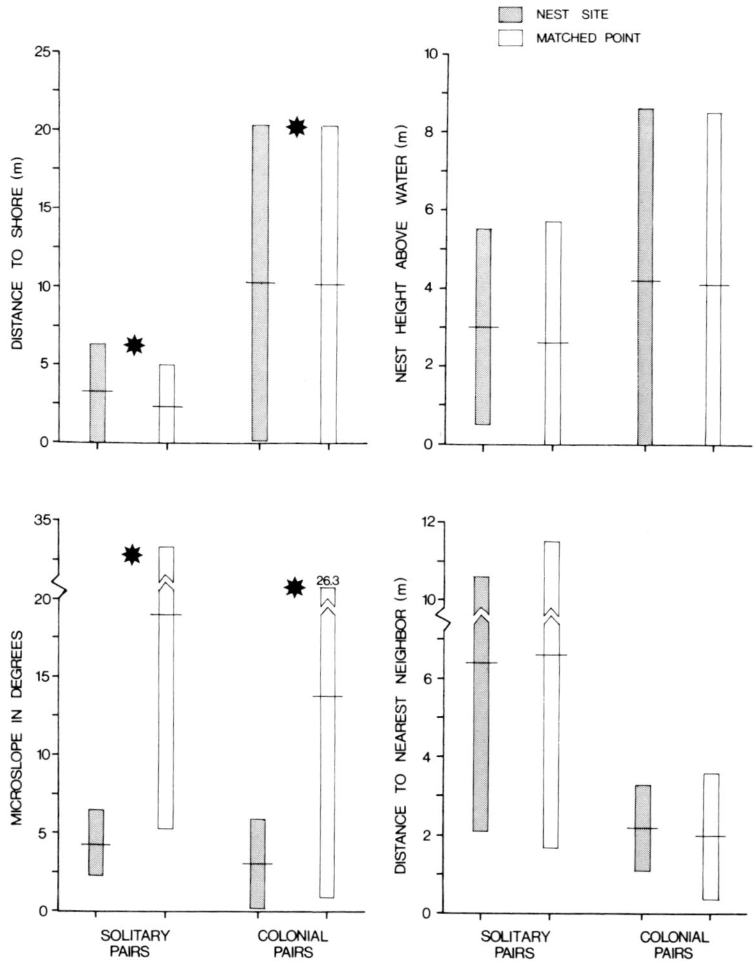

FIGURE 8. Means and 95 percent confidence intervals of variables for nest sites and their matched points of 12 solitary and 42 colonial pairs of Glaucous-winged Gulls in the Gulf Islands, 1985. An asterisk indicates significant difference ($P < 0.05$) by normal approximation of the Sign Test.

winged Gull, like the Mew Gull, is very plastic in its choice of nesting habitat. The recent invasion by Glaucous-winged Gulls of inland lakes and rivers, albeit on a limited scale, and on roofs of city buildings may result from a rapidly expanding population, facilitated by an increasing supply of human refuse (Drent and Guiguet 1961, Vermeer 1963, Butler et al. 1980).

ACKNOWLEDGMENTS

The authors thank L. Rankin and L. Willis for assistance with observations on Mew Gulls. This study was a research project of Environment Canada, Canadian Wildlife Service, Pacific and Yukon Region. It received additional assistance from Canada Department of Fisheries and Oceans, Institute of Ocean Sciences. R. W. Butler reviewed a preliminary draft. J. Hand, M. Gochfeld, W. E. Southern and an anonymous referee commented on the final draft of this manuscript and made pertinent comments. J. Balkwill and C. McKean of the B.C. Ministry of Environment provided access to information on lake inventories and nutrients, respectively.

LITERATURE CITED

ANDERSSON, M., F. GÖTMARK, AND C. G. WIKLUND. 1981. Food information in the Black-headed Gull, *Larus ridibundus.* Behav. Ecol. Sociobiol. 9:199–202.

BENT, A. C. 1921. Life histories of North American gulls and terns. U.S. Nat. Mus. Bull. 133. 345 pp.

BIANKI, V. V. 1967. Gulls, shorebirds and alcids of Kandalaksha Bay. Israel program for scientific translations, Jerusalem, 1977. 250 pp.

BURGER, J. 1981. A model for the evolution of mixed-species colonies of Ciconiiformes. Quart. Review Biology 56:143–167.

BURGER, J., AND E. SHISLER. 1980. The process of

colony formation among Herring Gulls *Larus argentatus* nesting in New Jersey. Ibis 122:15–26.

BURGER, J., AND M. GOCHFELD. 1981. Nest site selection by Kelp Gulls in southern Africa. Condor 83: 243–251.

BUTLER, R. W., N. A. M. VERBEEK, AND R. G. FOOTTIT. 1980. Mortality and dispersal of the Glaucous-winged Gulls of southern British Columbia. Can. Field Nat. 94:315–320.

CAMPBELL, R. W. 1970. Recent information on nesting colonies of Mew Gulls on Kennedy Lake, Vancouver Island, British Columbia. Syesis 3:5–14.

CAMPBELL, R. W. 1975. Marginal habitat used by Glaucous-winged Gulls for nesting. Syesis 8:395–396.

CONOVER, M. R. 1984. Inland breeding by the Glaucous-winged Gull. J. Field. Ornithol. 55:380–382.

DICK, M., I. M. WARNER, AND R. MACINTOSH. 1976. Notes on the biology of breeding and other seabirds in Chiniak and southern Marmot Bays, Kodiak Islands, Alaska, 1975. Adm. Rep. U.S. Fish & Wildl. Serv. OBS-CE, Anchorage, Alaska. 18 pp.

DRENT, R. H., AND C. J. GUIGUET. 1961. A catalogue of British Columbia sea-bird colonies. British Columbia Prov. Mus. Occ. Pap. No. 12. Victoria, British Columbia.

EDDY, G. 1982. Glaucous-winged Gulls nesting on buildings in Seattle. Murrelet 63:27–29.

FOOTTIT, R. G., AND R. W. BUTLER. 1977. Predation on nesting Glaucous-winged Gulls by river otters. Can. Field. Nat. 9:189–190.

GODFREY, W. E. 1966. The birds of Canada. Nat. Mus. Can. Bull. 203. 428 pp.

GÖTMARK, F. 1982. Coloniality in five *Larus* gulls: a comparative study. Ornis Scand. 13:211–224.

GÖTMARK, F., AND M. ANDERSSON. 1980. Breeding association between Common Gull *Larus canus* and Arctic Skua *Stercorarius parasiticus*. Ornis Scand. 11:121–124.

GÖTMARK, F., AND M. ANDERSSON. 1984. Colonial breeding reduces nest predation in the Common Gull (*Larus canus*). Anim. Behav. 32:485–492.

HAFTORN, S. 1971. Norges fugler. Universitetsforlaget, Oslo.

HANSSEN, O. J. 1984. Habitat selection of shorebirds in an archipelago in SE Norway. Ornis Scand. 15: 253–260.

HATCH, S. A., D. R. NYSEWANDER, A. R. DEGANGE, M. R. PETERSEN, R. A. BAIRD, K. D. WOHL, AND J. C. LENSINK. 1978. Population dynamics and trophic relationships of marine birds in the Gulf of Alaska and southern Bering Sea. Pp. 1–68 *in* Environmental Assessment of the Alaskan Continental Shelf. Ann. Rep. March 1978. Vol. 3. Birds. NOAA, Envir. Res. Lab. Boulder, Colorado.

HUMMEL, T. J. AND J. R. SLIGO. 1971. Empirical comparison of univariate and multivariate analysis of variance procedures. Psych. Bull. 76:49–57.

HUNT, G. L., JR., AND M. W. HUNT. 1976. Gull chick survival: the significance of growth rates, timing of breeding and territory size. Ecology 57:62–75.

MERILEES, W. J. 1974. A Glaucous-winged Gull mated to a Herring Gull on Okanagan Lake, British Columbia. Can. Field Nat. 88:485–486.

OLDAKER, R. F. 1963. Unusual nest site of the Glaucous-winged Gull. Can. Field Nat. 77:65–66.

SILVOLA, T. 1957. Kalakokki, *Larus canus L.,* Rakenta-nut pesansa ramemantyyn. Orn. Fenn. 34. No. 4.

SOWLS, A. L., S. A. HATCH, AND C. J. LENSINK. 1978. Catalog of Alaskan seabird colonies. U.S. Dept. Int. Fish and Wild. Serv. Biol. Serv. Prog. FWS/OBS-78/78.

STRANG, G. L. 1974. Nesting and feeding ecology of Mew Gulls (*Larus canus*) and Sabine's Gulls (*Xema sabini*) on the Kashunuk River. Prog. Rep. Dept. Forestry and Nat. Resources. Purdue University. West Lafayette, Indiana, 10 pp.

VAISANEN, R. A., AND O. JARVINEN. 1977. Quantitative structure and primary succession of bird communities in a Finnish archipelago. Ornis Scand. 8: 47–60.

VERBEEK, N. A. M., AND J. L. MORGAN. 1978. River Otter predation on Glaucous-winged Gulls on Mandarte Island, British Columbia. Murrelet 59:92–95.

VERMEER, K. 1963. The breeding ecology of the Glaucous-winged Gull *Larus glaucescens* on Mandarte Island, B.C. Occ. Pap. British Columbia Prov. Mus. 13:1–104.

VERMEER, K. 1970. Breeding biology of California and Ring-billed Gulls: a study of ecological adaptation to the inland habitat. Can. Wildl. Serv. Rep. Ser. 12:1–52.

VERMEER, K. 1973. Some aspects of the requirements of Common Loons in Alberta. Wilson Bull. 85:429–435.

VERMEER, K., AND K. DEVITO. 1986. The nesting biology of Mew Gulls (*Larus canus*), on Kennedy Lake, British Columbia, comparison with Mew Gulls in northern Europe. Colonial Waterbirds 9:95–103.

VOOUS, K. H. 1960. Atlas van Europese vogels. Elsevier, Amsterdam. 284 pp.

WARD, J. G. 1973. Reproductive success, food supply, and the evolution of clutch size in the Glaucous-winged Gull. Ph.D. thesis, Univ. British Columbia. 119 pp.

WESELOWSKI, T., E. GLAZEWSKA, L. GLAZEWSKI, B. NAWROCKA, P. NAWROCKI, AND K. OKONSKA. 1984. Distribution and numbers of waders, gulls and terns nesting on islands in the middle course of the Vistua. Acta. Orn. 20:159–185.

WILLIAMSON, F. S. L., AND L. J. PEYTON. 1963. Interbreeding of Glaucous-winged and Herring Gulls in the Cook Inlet region, Alaska. Condor 65:24–28.

YTREBERG, N. J. 1956. Contribution to the breeding biology of the Black-headed Gull (*Larus ridibundus* L.) in Norway. Nest, eggs and incubation. Nytt. Mag. Zool. 4:5–106.

ZAR, J. H. 1974. Biostatistical analysis. Prentice-Hall. Englewood Cliffs, N.J. 620 pp.

Studies in Avian Biology No. 10:119-128, 1987.

BEHAVIORAL CONSEQUENCES OF HABITAT SELECTION IN THE HERRING GULL

RAYMOND PIEROTTI[1]

ABSTRACT.—Data were collected on time budgets, rates of chick provisioning, and patterns of aggressive behavior in Herring Gulls *Larus argentatus* that nested in three distinct habitats on Great Island, Newfoundland. Exposed marine terraces (rocky habitat) had the highest nest density, and territories were subject to high levels of intrusion by prospecting conspecifics. This resulted in high levels of aggressive interaction, yet these birds had high breeding success. In contrast, gulls nesting in meadows suffered a high rate of predation on eggs and young. This predation pressure forced male birds to remain on their territories to defend their nests, which resulted in high rates of neighbor-neighbor aggressive interaction, and a reduced rate of chick provisioning by males. The third habitat, grass-hummock covered maritime slopes (puffin habitat) had low nest density and little or no predation pressure. This resulted in low levels of aggressive interaction and reduced vigilance with no apparent decline in offspring production. The results of this study demonstrate how habitat choice can have behavioral consequences that contribute to variation in offspring production within a species.

Most investigations of habitat selection in birds stress comparisons between species (e.g., Klopfer 1963, Hilden 1965; Partridge 1974, 1976, 1978), or the adaptation of species to specific types of habitat (e.g., Cullen 1957). These studies emphasize specific adaptations to particular habitats and how these adaptations allow different species to coexist (e.g., Cody 1985). Another important, but often ignored, aspect of avian ecology is variation within one species in the selection of a habitat for nesting, and the costs and benefits of settling in one habitat as opposed to another, since selection pressures—e.g., intensity of predation, availability of food—may vary among habitats. Investigations of these costs and benefits provide insights into how natural selection works by demonstrating how fitness (measured as the number of surviving offspring) varies within a population.

Studies of seabird breeding biology have provided some of the best examples of this phenomenon. Nettleship (1972) demonstrated variation in reproductive output over two seasons between habitats in the Atlantic Puffin (*Fratercula arctica*). Burger (1984) has shown that nest density and reproductive output varied among habitats in Herring Gulls (*Larus argentatus*) in New Jersey, although in this study data were lumped from different habitats over a three-year period. Finally, Pierotti (1982) showed that laying and hatching dates, clutch size, egg weight, hatching success, chick growth rates, and fledging success varied consistently among habitats and between years in the Herring Gull in Newfoundland.

To effectively demonstrate the costs and benefits of settling in a particular habitat, it is necessary to collect data on the behavior, ecology, and reproductive output of different individuals of the same species and sub-populations that live in neighboring but different habitats. Observations of the amount of time spent in various activities—e.g., anti-predator activities or foraging—by different individuals in different habitats allows the assessment of the magnitude of various selection pressures and the possible impact of interactions between these factors. Data have been published on breeding biology of Herring Gulls in three habitats on Great Island, Newfoundland (Pierotti 1982). In this paper, data collected on activity budgets, rates of chick provisioning, the nature and rates of aggressive interaction, and the impact of predation and intraspecific competition on these factors in Herring Gulls in the same three habitats will be described.

METHODS

This study was carried out between 10 June and 10 August 1976, and 5 May through 10 August in 1977 and 1978 on Great Island, Newfoundland (47°11′N, 52°46′W). For general descriptions of the island see Nettleship (1972) and Pierotti (1982). Data were collected in three distinct habitats, which were designated as "rocky" (exposed marine terraces at the southern and western periphery of the island), "puffin" (maritime slopes covered with grassy hummocks where Atlantic Puffins nested in large numbers), and "meadow" (flat, grassy areas on flat plateaus atop the southern portion of the island). Data were collected daily on breeding biology (laying and hatching dates, egg and clutch size) from 5 May until 1 June, and then on alternate days from 1 June until 31 July (hatching dates, chick growth and mortality; see Pierotti 1982).

In 1977 and 1978, continuous, all day (0600–2100) observations of behavior were conducted from blinds located in each of the three habitats on a rotating basis from the first week of June until August. These all-day watches were alternated with days during which data on breeding biology were collected, so that three or four days a week were spent observing behavior until the end of July. Observations were conducted using a zoom (20–60×) spotting scope or 10 × 50 binoculars.

To minimize disturbance, an effort was made to recognize individuals without handling adult birds. Some birds were identified using a system in which distinct

[1] Department of Biological Sciences, University of California, Santa Barbara, California 93106.

TABLE 1
SIZES OF MEALS FED TO HERRING GULL CHICKS ON GREAT ISLAND, NEWFOUNDLAND

Year	Habitat	Capelin per meal Mean ± SD (N)	Squid per meal Mean ± SD (N)
1977	Rocky	3.6 ± 0.8 (112)	2.2 ± 0.3 (61)
	Puffin	3.5 ± 1.0 (63)	2.0 ± 0.4 (35)
	Meadow	3.5 ± 0.7 (87)	2.1 ± 0.4 (29)
1978	Rocky	4.2 ± 1.1 (73)	2.3 ± 0.6 (24)
	Puffin	3.8 ± 1.3 (62)	2.1 ± 0.8 (16)
	Meadow	3.9 ± 1.0 (84)	2.2 ± 0.8 (29)

natural markings (e.g., holes or tears in webbing of feet, scratches on bill) were noted (Pierotti 1981). Males and females were identified by noting the role taken in copulation, by natural markings, and by the larger size of the male (Fox et al. 1981, Pierotti 1981). Many individuals were also dyed for identification using Rhodamine (puffin habitat), Malachite green (meadow habitat), and picric acid (rocky habitat). The technique employed was to remove the entire clutch from a nest. A piece of plastic wrap and a dummy clutch (of hard-boiled gull eggs) were then placed in the nest cup, and dye was poured over these eggs. When birds returned to the nest they marked themselves by sitting on and shifting the eggs. After a period of one to two hours, the dummy eggs and plastic wrap were removed and the original clutch was returned. This procedure appeared to have no effect on subsequent hatching or fledging success.

In each of the three habitats, twelve territories on which both the male and female were individually recognizable were selected for monitoring during the 1977 breeding season. The same 12 pairs were monitored in rocky habitat during both 1977 and 1978. In puffin habitat 10 of the original 12 pairs, and in meadow habitat 11 of the original 12 pairs were monitored in both years. These individuals were recognizable from one year to the next by individual markings. The same number of days were spent observing birds in each habitat.

Data on time budgets were collected using a combination of scan-sampling and ad libitum sampling (Altmann 1974) on the 12 focal territories. Every 15 minutes all twelve nests were scanned and the presence or absence of individuals and the activity in which they were engaged, if present—e.g., incubating, sleeping—was noted. Data on discrete behavioral acts (e.g., chick feeding, aggressive acts) were recorded during the intervals between scans. During observation periods, the type and number of food items presented to chicks in a given feeding were noted whenever possible. This was not difficult, because Herring Gulls typically regurgitated entire freshly caught fish or squid. All such data were collected by observation through the spotting scope.

Acts of aggressive behavior were noted as to type, identity of aggressor, and identity of recipient of the aggression (e.g., neighbor male, intruder female; all neighbors were individually recognizable). Eight distinct aggressive displays were noted and subdivided into three levels of aggressive intensity that reflect the possible energetic cost and risk of the display. Aggressive uprights, long-calls, and choking (see Tinbergen 1960) were simple displays and considered to be of low intensity (Level 1). More active and energetically costly displays, e.g., grass-pulling and supplants (running at an individual and causing it to take flight), were considered to be of moderate intensity (Level 2). Interactions involving active pursuit or physical contact, e.g., fights, hitting, supplants followed by aerial pursuit, were considered to be of high intensity (Level 3). Long calls and choking may occur in both aggressive and sexual contexts (Hand 1979); therefore, the contexts of displays were noted, and those acts performed between members of a pair were not included in the analysis of aggressive behavior.

I also noted the number of adult gulls other than neighbors that intruded into territories during 15 hours of observation in each habitat in 1978 during which no other data were collected. These birds were clearly not residents of the area since they were not recognizable, and were probably non-breeding birds scouting for available space for nesting. Using a stopwatch, I timed the interval from the time one of these "intruders" landed until it was evicted by a resident individual.

All data sets were tested for normality and homoscedasticity. Where these conditions were met, parametric statistics (t-test, ANOVA) were employed. In cases where distributions were either non-normal or excessively heteroscedastic, other statistical tests were employed, e.g., Chi-square, Wilcoxon's signed-ranks test, or Kruskal-Wallis test. In cases where ANOVAs were found to show significant differences among samples, Student-Newman-Keuls (SNK) tests were used to examine range differences among samples (Sokal and Rohlf 1981).

RESULTS

TIME BUDGETS

The 1977 reproductive period was very stormy early in the breeding season and food appeared to be difficult to obtain because (1) more Herring Gulls were observed to feed on garbage, a food on which they did more poorly (31% of population specialized on garbage) early in the breeding season than in 1978 (21% garbage specialists; Pierotti and Annett 1986), and (2) meal sizes of capelin (*Mallotus villosus*: a principal food fish of seabirds on Great Island) were smaller than in 1978 (Table 1). In 1977, capelin did not appear in large numbers in the waters around the island (Pierotti and Annett 1986). As a result, egg-laying was delayed, egg and clutch sizes were smaller, and chicks grew at a slower rate than during the 1978 breeding season (Pierotti 1982).

The 1978 breeding season was very calm compared with 1977, and food appeared to be abundant. Garbage declined in the diet (Pierotti and Annett 1986). Capelin were abundant in the waters around Great Island beginning in early June as indicated by the increase in meal size, the foraging patterns of the gulls, and the presence in the area of large numbers of humpback whales, *Megaptera novaeangliae,* which also feed on capelin. Egg size, clutch size, and chick growth rates all increased significantly compared with 1977 (Pierotti 1982).

Incubation period

I define the incubation period as the period from the laying until the hatching of the first-laid egg. During this period, one member of each pair was on the nest except when disturbed, or during brief (1–2 minute) absences to defecate or drive off an intruder.

During the 1977 breeding season female Herring Gulls nesting in puffin habitat were present on their territories significantly less time per day than female gulls in either rocky or meadow habitat (mean = 64 ± 16.3 minutes, $P < 0.05$ by ANOVA). Male and female gulls spent equivalent amounts of time present on their territories in puffin and meadow habitat, whereas male gulls in rocky habitat were present significantly less than their mates (Fig. 1a). Overall, puffin habitat had the lowest percentage of time when both the male and female were present together on the territory, whereas meadow habitat had the highest percentage of time when both members of the pair were present.

The role of male and female gulls in incubation in the three habitats reveals a similar pattern. There were significant differences in the amount of time spent in incubation between the sexes in all three habitats, with females spending more time incubating. This difference was smallest in puffin habitat where males spent significantly more time in incubation than males in other habitats (Fig. 1a). In contrast, male gulls in rocky habitat spent the least time in incubation. Male gulls in rocky habitat spent only 51% of their total time on the territory incubating compared with 70% for males in puffin habitat.

Time budgets were quite different in rocky and meadow habitats during the 1978 breeding season. As in 1977, female gulls nesting in puffin habitat were present less (mean = 47 ± 14.7 minutes, $P < 0.05$ by ANOVA), and males and females spent less time together on the territory than in the other habitats. There was, however, no significant difference between males and females in amount of time spent incubating in puffin habitat in 1978. In meadow habitat females spent more time than males on the territory, which was not the case in 1977, and in incubation (cf. Fig. 1a and 1b). Also in contrast to 1977, there was no difference in the amount of time male and female gulls in rocky habitat were present, although males spent slightly less time incubating. Male gulls in meadow habitat, however, were present significantly more in 1977 than in 1978.

This variation in attendance of male gulls in meadow habitat appeared to be related to intensity of nest predation. During all three years of this study, the number of eggs and chicks that disappeared was greatest in meadow habitat (Table 2). Most of these disappearances were probably attributable to predation both by conspecifics and by Great Black-backed Gulls, *Larus marinus,* which nested almost exclusively in meadow habitat. In 1977, when food supplies were low, percentages of egg and chick disappearance were higher than in any other year (Table 2). This increase in nest predation was most marked in meadow habitat where nearly 20% of the eggs and 35% of the chicks produced by the 90 pairs monitored in this habitat disappeared. During such periods, male vigilance increased, as indicated by the increased amount of time spent on the territory by male gulls in meadow habitat, and by the increased rate of aggressive interaction shown by males in this habitat during 1977 (see below).

Chick period

For each nest, the chick period was defined as the period from when the first chick hatched until the last surviving chick fledged from the nest. Chicks were considered fledged when they were capable of sustained, independent flight away from the natal territory, Typically at least one parent was in attendance on the territory at all times during this period.

In contrast to the incubation period, there were no significant differences in the amounts of time spent on the territory by males and females in any of the three habitats during either year (Fig. 1a, b). There were also no significant differences in attendance patterns among the habitats, since either the male or female was present on the territory at all times.

During the 1977 season broods of chicks in puffin habitat were fed significantly less frequently than broods in the other two habitats (Table 3). During both years there were significant differences in feeding rate between male and female gulls in rocky habitat, and the rates of chick feeding by males and females were identical in meadow habitat (Table 3). Despite the difference in feeding rates, there were no significant differences in meal sizes among the habitats for either capelin or squid (*Illex illecebrosus*; Table 1). These two food types constituted over 95% of all chick feedings observed.

There was no obvious relationship between feeding rates and chick survival (Table 3). During 1977 in territories monitored during behavioral observations, fledging success was highest in rocky habitat, followed by puffin habitat. Meadow habitat had the lowest fledging success ($P < 0.01$ by

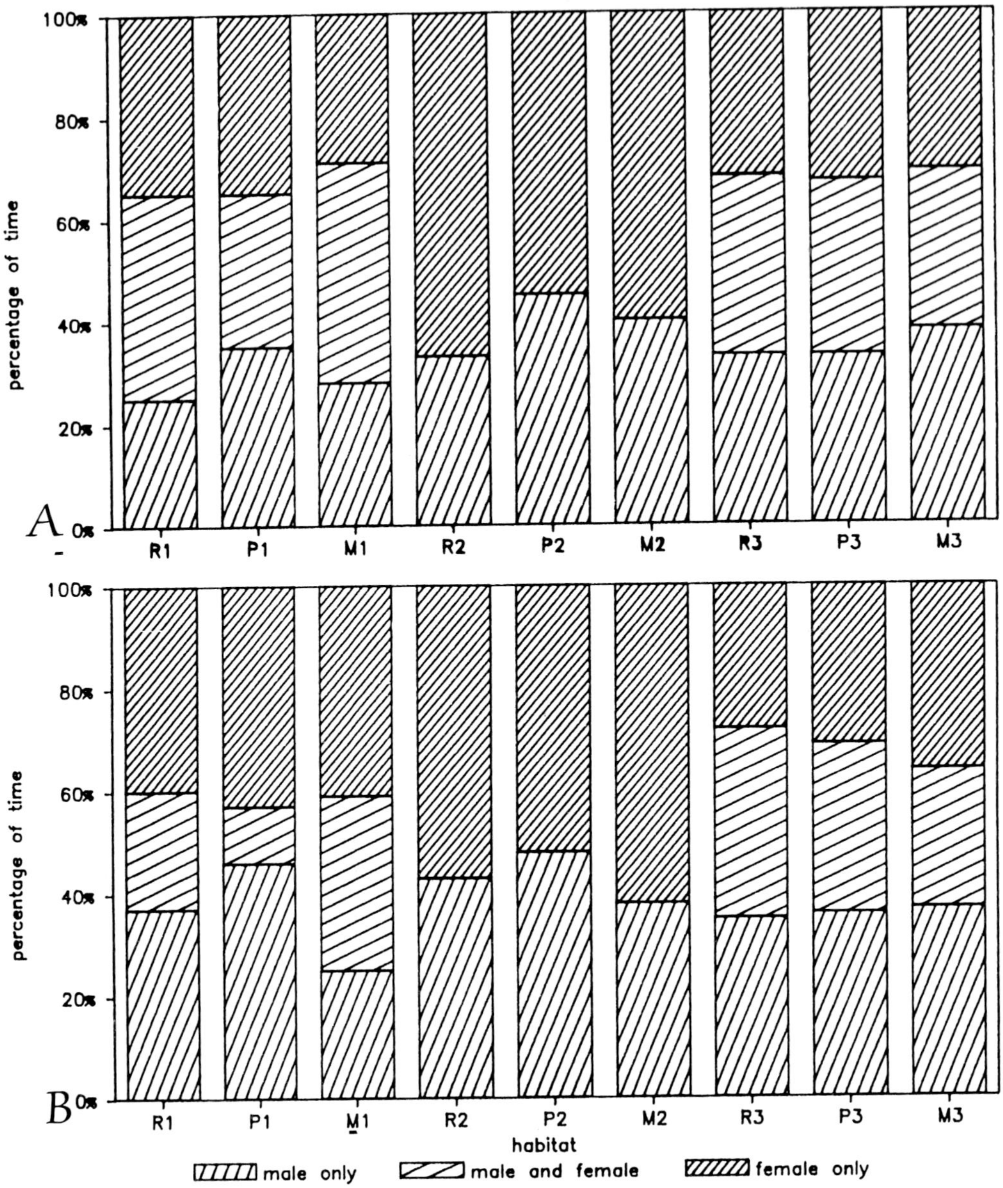

FIGURE 1. Activity budgets of male and female gulls during 1977 and 1978 breeding seasons. Data expressed as percentage of total time spent in observation = 15 hours/day over 6 days in both years. R1, P1, and M1 = birds present in rocky (R), puffin (P) and meadow (M) habitats, respectively, during May and early June (incubation period); R2, P2, and M2 = birds on nests incubating; R3, P3, M3 = birds present on territories during late June and July (chick-rearing).

A. 1977 season: Amount of time spent on territory during incubation differs among habitats for females and for males and females together at 0.5 level by ANOVA (for 1's, P less than R and M at .01 level by SNK). Males incubate significantly less than females in all three habitats (paired sample t test; for R, $P < .001$, for P, $P < .05$, and for M, $P < .01$).

B. 1978 season: Results similar except that males incubated significantly less than females only in R and M (paired sample t test; for R, $P < .05$, and for M, $P < .01$).

TABLE 2
PERCENTAGES OF HERRING GULL EGGS AND CHICKS THAT EITHER DISAPPEARED OR WERE EATEN DURING THREE YEARS ON GREAT ISLAND, NEWFOUNDLAND

Year	Habitat: Rocky	Puffin	Meadow	% Killed by GBBG (meadow habitat only)
A. Eggs[1]				
1976[3]	0.6 (161)[2]	3.2 (155)	5.9 (188)	not applicable
1977[4]	9.4 (286)	11.6 (301)	18.9 (228)	n.a.
1978[4]	8.6 (304)	7.7 (363)	13.0 (223)	n.a.
B. Chicks				
1976[4]	5.5 (121)	21.3 (122)	31.5 (143)	9.8
1977[4]	7.6 (131)	16.7 (138)	34.2 (111)	9.9
1978[4]	2.1 (141)	9.4 (202)	17.1 (111)	14.4

[1] Data in 1976 incomplete because of arrival on island after peak of clutch completion.
[2] (N) = total number monitored in each habitat
[3] $P < .05$ by Chi Square test (among habitats).
[4] $P < .001$ by Chi Square test (among habitats).

chi square, 4 df) despite the fact that chicks in this habitat were provisioned at the same rate as in rocky habitat. This pattern was similar to the pattern observed for all nests monitored on Great Island in 1977, except that rocky and puffin habitat had similar rates of fledging success (Pierotti 1982). In 1978, there were no significant differences in fledging success, although the pattern observed in provisioning rate among the three habitats for the 12 pairs monitored during behavioral observations was the same as in 1977 (Table 3). For all nests monitored during 1978 there were no significant differences in fledging rates between rocky and puffin habitats but pairs in meadow habitat had significantly lower fledging success (Pierotti 1982). In both years, however, chicks in rocky habitat grew at significantly faster rates than did chicks in the other two habitats (Pierotti 1982, tables 5–7).

AGGRESSIVE BEHAVIOR

Patterns of aggressive behavior in both male and female gulls varied among habitats. This variation could be related to differences in time budgets, fledging success, and the three-dimensional structure of the habitats. Rocky habitat

TABLE 3
RATE OF BROOD PROVISIONING IN RELATION TO BREEDING SUCCESS IN ALL THREE HABITATS ON GREAT ISLAND DURING THE BREEDING SEASONS OF 1977 AND 1978

Habitat	Sex	Chick feedings/hr[a,b] (mean ± SD)	Chicks fledged/pair (mean ± SD)
A. 1977			
Rocky (N = 12)	Male	0.19 ± 0.04[c,d]	
	Female	0.12 ± 0.02	2.00 ± 0.68[e]
Puffin (N = 12)	Male	0.13 ± 0.03	
	Female	0.10 ± 0.03	1.67 ± 0.65
Meadow (N = 12)	Male	0.15 ± 0.04	
	Female	0.15 ± 0.04	1.08 ± 0.79
B. 1978			
Rocky (N = 12)	Male	0.18 ± 0.05[d]	
	Female	0.11 ± 0.03	1.68 ± 1.01
Puffin (N = 12)	Male	0.14 ± 0.04	
	Female	0.10 ± 0.04	1.75 ± 0.83
Meadow (N = 12)	Male	0.14 ± 0.04	
	Female	0.14 ± 0.05	1.77 ± 0.89

[a] Chick feedings/hr represent the number of times an adult male or female gull returned to the territory after an absence greater than 20 minutes and fed chicks. It does not include repeat feeds from the same feeding bout.
[b] For both years, the total rate of chick provisioning (male + female) was significantly different among habitats at the .05 level by ANOVA ($P <$ R, M at .05 level by SNK).
[c] For rate of male provisioning, differences among habitats significant at .05 level by ANOVA (R greater than P, M at .05 level by SNK).
[d] In rocky habitat in both years, difference in provisioning rate by male and female gulls significant at .05 level by paired sample t test.
[e] In 1977, difference among habitats in fledging rate significant at .05 level by Chi Square test, with 4 df.

TABLE 4

FREQUENCY OF AGGRESSIVE INTERACTIONS (NUMBER PER HOUR) OF MALE AND FEMALE HERRING GULLS ON GREAT ISLAND DURING THE REPRODUCTIVE SEASONS OF 1977 AND 1978

Habitat (N = no. pairs)	Sex	Interactions between neighbors[a] Level 1	Level 2	Level 3	Total between neighbors	Interactions between residents and intruders Level 1	Level 2	Level 3	Total between resident and intruder
A. 1977									
Rocky (12)	Male[b]	1.45	1.85	0.56	3.86	0.74	2.91	1.23	4.88
	Female[c]	0.72	0.24	0.07	1.03	0.57	1.43	0.29	2.29
Puffin (12)	Male	1.00	3.38	0.78	5.16	0.28	1.05	0.41	1.74
	Female	0.32	0.13	0.04	0.49	0.13	0.53	0.76	1.42
Meadow (12)	Male	1.88	5.74	1.08	8.70	0.51	1.39	0.25	2.15
	Female	0.85	0.43	0.12	1.40	0.30	0.46	0.08	0.84
B. 1978									
Rocky (12)	Male[b]	1.70	1.84	0.41	3.95	1.35	2.49	1.00	4.84
	Female[c]	1.10	0.30	0.10	1.50	0.64	0.94	0.24	1.82
Puffin (12)	Male	0.97	2.45	0.20	3.62	0.26	0.44	0.27	0.97
	Female	0.46	0.17	0.01	0.64	0.13	0.31	0.04	0.48
Meadow (12)	Male	2.17	3.41	0.63	6.21	0.64	0.66	0.23	1.53
	Female	1.00	0.99	0.31	2.30	0.21	0.31	0.24	0.76

[a] In meadow habitat neighbors include Great Black-backed Gulls.
[b] Rates of aggression among habitats significantly different for males at .001 level by Chi Square test, 5 df.
[c] Rates of aggression among habitats significantly different for females at .01 level by Chi Square test, 5 df.

was most heterogeneous, with the highest density of nesting birds (Pierotti 1982). Puffin habitat was the least spatially heterogeneous with the lowest nesting density and meadow habitat was intermediate in both variables.

During both years of the study, male and female gulls in rocky habitat directed more aggression towards nonresident conspecifics (intruders) than towards neighbors (Table 4; $P < 0.001$, Fisher's exact test). The rate of intrusion in rocky habitat was more than twice the rate of intrusion in either of the other habitats (Table 5). During both 1977 and 1978, female gulls in rocky habitat directed more aggression at intruders than did male gulls in either puffin or meadow habitat (Table 4). These intruders were not "loafers," seeking temporary resting sites. Nearly all were adult birds and appeared to be seeking nesting sites. When not displaced by resident birds, they were often joined by a partner of the opposite sex, with whom they would participate in choking displays and initiate nest construction. Other intruders not displaced by resident birds would approach residents while producing vocalizations typical of pair-formation or bonding, e.g., begging calls or mew calls (Tinbergen 1960, Hand 1979). As a result, intruders elicited responses of moderate and high intensity at a significantly

TABLE 5

INTRUDER DISPLACEMENT BEHAVIOR OBSERVED DURING 1978 BREEDING SEASON (EACH HABITAT OBSERVED FOR 15 HOURS)

Resident-intruder	Time until intruder displaced (mean ± SD in seconds) Rocky habitat (N)	Puffin habitat[a] (N)	Meadow habitat[a] (N)
Male vs. Male[b,c]	11.3 ± 15.9 (54)[d]	36.7 ± 43.7 (10)	35.6 ± 46.2 (15)
Male vs. Female[c]	15.4 ± 19.7 (36)[d]	34.8 ± 29.4 (10)	99.7 ± 82.9 (23)
Male vs. Intruder (total)[c]	12.9 ± 16.2 (90)[d]	35.8 ± 36.5 (20)	60.4 ± 66.5 (38)
Female vs. Male	21.1 ± 20.3 (15)	None observed	18.2 (1)
Female vs. Female[c]	17.3 ± 21.1 (15)	22.9 ± 29.2 (4)	62.9 ± 88.5 (11)
Female vs. Intruder (total)[c]	19.7 ± 20.4 (30)[d]	33.9 ± 29.2 (4)	59.1 ± 83.2 (12)

[a] Intruders that remained for at least 300 seconds (5 minutes) before being displaced are not included in calculations (there were 12 such birds in meadow habitat and 7 in puffin habitat, none was observed in Rocky).
[b] Male vs. Male = Male resident displacing male intruder, etc.
[c] Difference in duration among habitats significant among habitats at .001 level by Kruskal Wallis test.
[d] Difference in number of intruders significant at .01 level by Chi Square test.

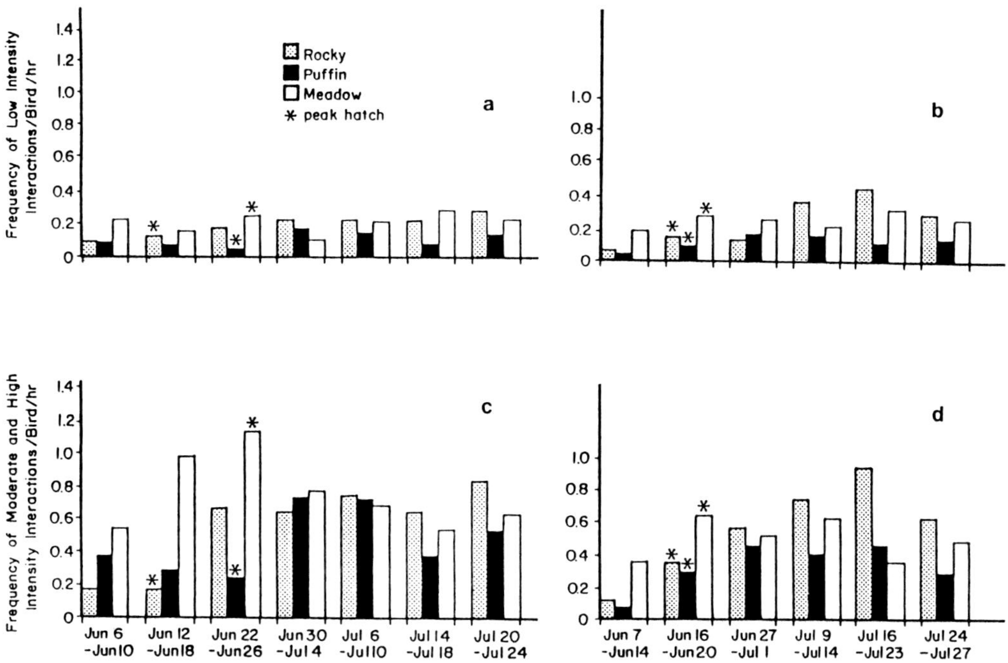

FIGURE 2. Temporal pattern of aggressive behavior by male Herring Gulls during the breeding season. 2a,c: 1977; 2b,d: 1978. In both years at all levels of intensity, patterns of aggression among habitats are significantly different at the .001 level by χ^2 contingency test with 12 df.

greater rate than did interactions between neighbors ($P < 0.001$ by chi square test, 2 df).

In contrast to rocky habitat, gulls in the other habitats typically directed aggression at a much higher rate at neighbors than at intruders (Table 4, $P < 0.001$, Fisher's exact test). During both 1977 and 1978, male and female Herring Gulls nesting in puffin habitat showed the lowest frequencies of aggressive interaction against both neighbors and intruders combined. Despite the low nest density in this habitat, however, rates of interaction between neighboring male gulls were higher in puffin habitat than in rocky habitat in 1977. Rates of interaction between neighbor males were similar in these two habitats in 1978 (Table 4). During both years of this study, gulls in meadow habitat suffered high rates of egg and chick loss (Table 2), and showed the highest rates of neighbor-neighbor aggression ($P < 0.001$ by Chi square, 4 df). The perpetrators of this aggression were resident male gulls that spent more time on their territories after losing eggs early in the season. In addition to guarding mates and young, these males also foraged in this habitat. Most of these males were specialist predators on Leach's Storm Petrel, *Oceanodroma leucorhoa,* which nested in large numbers in this habitat (Pierotti and Annett 1986). Some of these males also stole and ate eggs and young of neighboring conspecifics. This predation led to increased vigilance and aggression by their victims, and this pattern occurred repeatedly until nearly all birds in this habitat were aggressive towards neighbors. In 1977 when food was less abundant, both the percentage of egg and chick disappearance and rates of aggression were higher than in 1978 when food was abundant (cf. Tables 2 and 4).

These high rates of aggression directed at neighbors appeared to result initially from harassment and predation by Great Black-backed Gulls. In every case where a pair of Herring Gulls was observed to lose one or more chicks or eggs to predation in meadow habitat (n = 19), there was a significant increase in rates of aggression by both the male and female parents within 12 hours ($P < 0.001$ by Wilcoxon signed-ranks test). Similar increases in rates of aggression were observed in pairs in rocky and puffin habitats following loss of chicks (n = 11, $P < 0.01$ by Wilcoxon). Neither infanticidal behavior nor increased rates of aggression were detected in pairs that did not lose eggs or chicks.

Temporal changes in rates of aggressive behavior were found in all three habitats in relation to the stage of the breeding cycle. During the latter stages of incubation until the peak of hatching (6–22 June) in both 1977 and 1978, male

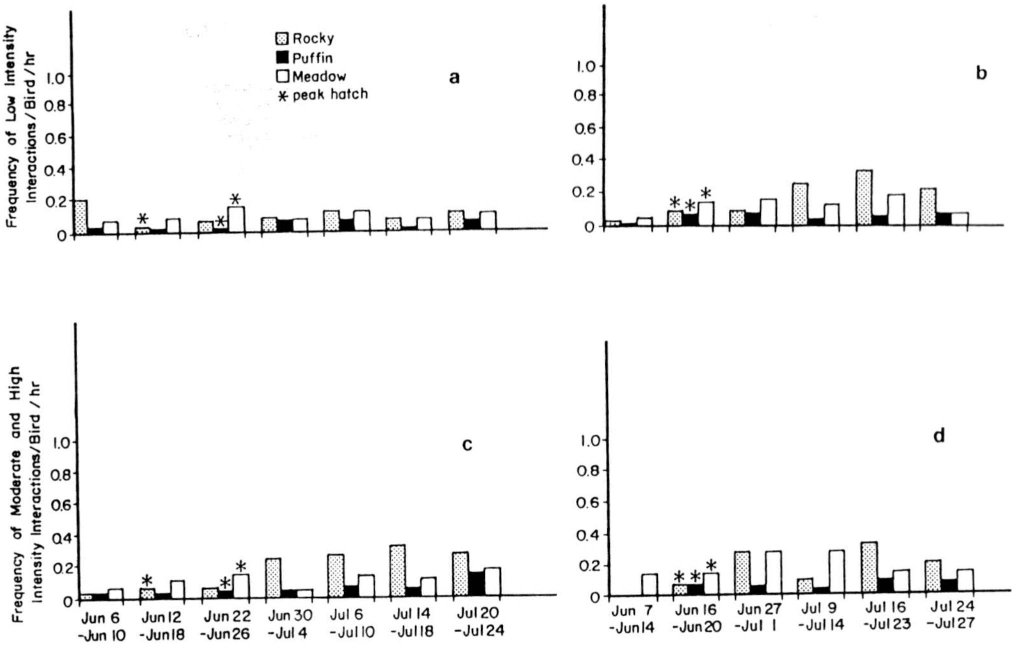

FIGURE 3. Temporal pattern of aggressive behavior by female Herring Gulls during the breeding season. 3a,c: 1977; 3b,d: 1978. In both years at all levels of intensity, patterns of aggression among habitats are significantly different at the .001 level by χ^2 contingency test with 12 df.

gulls in meadow habitat showed the highest rates of both low and high intensity aggressive behavior (Fig. 2). This was the result of increased aggressiveness by male gulls in this habitat after losing eggs or newly hatched chicks. Within a week after peak hatch, there was a marked increase in aggression by male gulls in rocky habitat, and from this point until the end of July in both years, male gulls in rocky habitat had the highest rates of aggressive interaction (Fig. 2). These birds were responding to frequent intrusions by prospecting birds, and to attacks on chicks by neighbors in this high density habitat, where chicks were more likely to cross boundaries than in other less dense habitats. A similar, but smaller scale increase in rate of aggression was observed to occur in puffin habitat.

Female gulls showed a pattern similar to that of their mates but with overall lower rates of aggression (Fig. 3). Rates of female aggression were highest in meadow habitat until peak hatching, and thereafter were highest in rocky habitat. Most aggression by female gulls in rocky habitat was directed at intruders or neighbors that approached chicks.

Intruders landing on occupied territories were observed far more frequently in rocky habitat and were displaced most rapidly in this habitat (Table 5). Female gulls in rocky habitat displaced intruders of both sexes more rapidly than did males in either of the other habitats. Adult intruders elicited the most rapid and intense responses, especially when they landed near chicks. In puffin habitat some intruders were able to establish territories, build nests, and lay eggs. In meadow habitat some intruders were able to remain long enough to initiate nest-building. In rocky habitat no intruder was observed to remain long enough to even attempt such behavior.

DISCUSSION

Settling in the three habitats appeared to impose varying costs on breeding Herring Gulls. Costs in terms of time and energy were probably highest in meadow habitat because of the need for increased vigilance to guard eggs and young. These costs took three forms: First, energy expenditure in aggression was great in this habitat because rates of aggression were high here during both years. Second, the need for vigilance probably reduced time available for foraging. This was especially true where foraging by males was concerned, since male gulls usually provide more food to offspring than females (Burger 1981, Pierotti 1981, Bellrose 1983). In meadow habitat male gulls provided less food than did males in rocky habitat. The final cost in this habitat was a reduction in fitness, as indicated by the fact that the fledging rate was lowest here during all three years of this study (Pierotti 1982).

In rocky habitat benefits appeared to exceed costs. The rocky marine terraces at the periphery of the island appeared to be the preferred habitat for nesting, and were occupied by phenotypically superior individuals (Pierotti 1982). Rocky habitat was the best because it was the most spatially heterogeneous habitat and contained numerous sites for nesting that were well sheltered from prevailing winds and storms (Pierotti 1982, tables 9, 10). In addition, this area was sparsely vegetated. During rain or heavy fog (which are common in Newfoundland during the spring and summer), the rocks that make up most of the habitat did not become saturated with moisture and dried quickly. In meadow and puffin habitats the abundant vegetation took up moisture, and the plumage of birds (especially downy chicks) remained soaked for hours or days afterward. These soaked chicks almost certainly lost heat at greater rates than did dry chicks, and so had less energy available for growth. Chicks in meadow habitat grew more slowly than chicks in rocky habitat during both years even though they were fed at identical rates (Table 3 and Pierotti 1982).

There were two apparent costs to nesting in rocky habitat, and both appeared to result from the apparent preference of Herring Gulls for settling in this habitat. The first cost was the consistently high rate of intrusion by unestablished adults which provoked aggressive responses of high and moderate intensity from residents. The second cost was the high nest density in this habitat and the resultant high level of aggression. Internest distances and territory sizes were significantly smaller in rocky habitat than in puffin and meadow habitat (Pierotti 1982, table 9). Since this habitat was heterogeneous, there was relatively little interaction between neighbors during the incubation period, possibly because as suggested by Burger (1977), these birds could not see each other while sitting on nests. After hatching, however, adults and chicks moved about their territories, some chicks were attacked, and some adults that lost chicks became egg or chick predators (see also Davis and Dunn 1976, Pierotti 1980). This combination of factors resulted in an increase in rate of aggressive behavior which began within one or two weeks after hatching and continued for the remainder of the breeding season.

The direct relationship between aggression and nest density is further supported by the data from puffin habitat which had the lowest density (Pierotti 1982, table 9), and the lowest rates of aggressive interaction during all phases of the breeding season in both years. The large distances between neighbors in this habitat apparently required little vigilance and reduced attendance on the territory during the incubation period. There appeared to be no cost to nesting in puffin habitat, since despite the lack of vigilance and the lowest rate of offspring provisioning in the three habitats, fledging success was as high or higher than in the other habitats. This situation apparently resulted from the absence of predatory Great Black-backed Gulls in this habitat, and the low preference for nesting in this habitat, which had significantly fewer nesting birds than were expected (Pierotti 1982, table 1). A reduction in rate of aggressive interaction under conditions of low nest density has also been reported in the Western Gull, *Larus occidentalis* (Ewald et al. 1980, Pierotti 1981).

Other investigations of gull breeding biology have demonstrated that increased nest density may lead to increased costs of reproduction. Schreiber et al. (1979) found that Laughing Gulls, *Larus atricilla,* laid equally large first and second eggs during the first year of their study. Nest density increased in the second year and although first-laid eggs remained similar in size to the previous year, there was a decrease in the size of second-laid eggs. Coulson et al. (1982) found that a reduction of nest density led to an increase in both adult body weight and egg weight in Herring Gulls. Finally, Burger (1984) showed an inverse relationship between fledging success and nest density in Herring Gulls in New Jersey that was apparently related to increased rates of egg predation and aggressive interaction at high nest density. My results demonstrate that nesting at high densities may lead to increased rates of aggressive interaction, which could reduce energy reserves necessary for egg production, or lead to reduced adult weight.

Although the results of this study and of Burger (1984) both suggest a cost to nesting at high density in Herring Gulls, our data actually reveal markedly different patterns. Burger concludes that fledging success is highest on intermediate-sized territories (internest distance = 6–7 m) and lowest in areas of high nest density (internest distance = 3–4 m; Burger 1984, table 35). The results of this study showed the highest rates of fledging success in areas of high (rocky; internest distance = 3–4 m) and low (puffin; internest distance = 8–9 m) nest density, and the lowest rate at intermediate nest density (meadow; internest distance = 6–7 m; Pierotti 1982, table 9).

Similarly, Burger concludes that rates of aggression in Herring Gulls are highest at high and low densities and lower at intermediate densities (1984, fig. 28b). She further argues that rate of intrusion by non-neighbors increases with increasing territory size (1984, fig. 29). My study demonstrates, however, that factors other than simple nest density may be responsible for these patterns. In all of these studies rates of aggression are highest where egg and chick production is greatest and fledging success is least. Burger (1984) notes this relationship, but characterizes it as "the most aggressive pairs fledged the fewest young (pg 74)," and presents data to show high levels of aggression at high and low nest densities. From my results it can be seen that birds that lose young become more aggressive, and that the habitat with the lowest density of nests also has the lowest rates of aggression, probably because intrusion rates are low and chicks are not attacked in this habitat.

Also in contrast to Burger (1984), in this study intrusion rate was greatest in the highest density (rocky) habitat. This habitat was preferred for nesting and therefore attracted the most intruders. Ewald et al. (1980) came to a similar conclusion after noting that small territories had the highest rates of intrusion in the Western Gull, and Coulson and Thomas (1983) noted that clutches were initiated earlier and rates of intrusion were higher in areas of high nest density in colonies of Kittiwakes, *Rissa tridactyla.*

These results suggest that the conclusions drawn by Burger (1984) are not of general applicability, and that the relationship between rate of aggression, territory size, and fledging success is best understood on a case-by-case basis. In the Herring Gulls on Great Island fledging success appears to be most strongly influenced by predation, and relatively independent of territory size. Similarly, rate of aggression appears to be independent of territory size, and the relationship between breeding success and aggression appears to take the form of an increase in aggression by birds that have lost offspring to predation or are under threat of predation. Therefore, predation on eggs and young, either by conspecifics or by other species, e.g., Great Black-backed Gulls, may be a key factor determining time budgets and reproductive output in Herring Gulls on Great Island.

ACKNOWLEDGMENTS

I thank C. Annett, R. G. B. Brown, M. Conover, M. K. McNicholl, I. A. McLaren, and N. Verbeek for comments on earlier drafts of this manuscript. Research funding and logistical support were provided by the Canadian Wildlife Service, and permission to work in the Witless Bay seabird sanctuary was granted by the provincial government of Newfoundland.

LITERATURE CITED

ALTMANN, J. 1974. Observational study of behaviour: sampling methods. Behaviour 49:227–267.

BELLROSE, C. A. 1983. The breeding biology of a small, mainland colony of Western Gulls. Unpublished M.A. thesis, San Jose State Univ.

BURGER, J. 1977. The role of visibility in the nesting behavior of five species of *Larus* gulls. J. Comp. Physiol. Psych. 9:1347–1358.

BURGER, J. 1981. On becoming independent in Herring Gulls: parent-young conflict. Amer. Nat. 117: 444–456.

BURGER, J. 1984. Pattern, mechanism, and adaptive significance of territoriality in Herring Gulls. Ornithol. Monogr. 34:1–92.

CODY, M. L. 1985. Habitat selection in birds. Academic Press, New York.

COULSON, J. C., N. DUNCAN, AND C. THOMAS. 1982. Changes in the breeding biology of the Herring Gull induced by reduction in size and density of the colony. J. Anim. Ecol. 51:739–756.

COULSON, J. C., AND C. S. THOMAS. 1983. Mate choice in the Kittiwake Gull. Pp. 361–376 *in* P. Bateson (ed.), Mate choice. Cambridge Univ. Press.

CULLEN, E. 1957. Adaptations in the Kittiwake to cliff-nesting. Ibis 99: 275–302.

DAVIS, J. W. F., AND E. K. DUNN. 1976. Intraspecific predation and colonial breeding in Lesser Black-backed Gulls. Ibis 118: 165–177.

EWALD, P. W., G. L. HUNT, JR., AND M. W. HUNT. 1980. Territory size in Western Gulls: importance of intrusion pressure, defense investments and vegetation structure. Ecology 61: 80–87.

FOX, G. A., C. R. COOPER, AND J. P. RYDER. 1981. Predicting the sex of Herring Gulls by using external measurements. J. Field Ornithol. 52: 1–9.

HAND, J. L. 1979. Vocal communication in the Western Gull, *Larus occidentalis*. Unpublished Ph.D. dissertation, Univ. California, Los Angeles.

HILDEN, O. 1965. Habitat selection in birds: a review. Ann. Zool. Fenn. 2: 53–75.

KLOPFER, P. R. 1963. Behavioral aspects of habitat selection: the role of early experience. Wilson Bulletin 75: 15–22.

NETTLESHIP, D. N. 1972. Breeding success of the Common Puffin (*Fratercula arctica*) on different habitats at Great Island, Newfoundland. Ecol. Monogr. 42: 239–268.

PARTRIDGE, L. 1974. Habitat selection in titmice. Nature 247: 573–574.

PARTRIDGE, L. 1976. Field and laboratory observations of the foraging and feeding techniques of Blue Tits and Coal Tits in relation to their habitat. Anim. Behav. 24: 534–544.

PARTRIDGE, L. 1978. Habitat selection. Pp. 351–370 *in* J. R. Krebs and N. B. Davies (eds.). Behavioural ecology. Sinauer Assoc., Sunderland, Mass.

PIEROTTI, R. 1980. Spite and altruism in gulls. Am. Nat. 115: 290–300.

PIEROTTI, R. 1981. Male and female parental roles in the Western Gull under different environmental conditions. Auk 98: 532–549.

PIEROTTI, R. 1982. Habitat selection and its effect on reproductive output in the Herring Gull in Newfoundland. Ecology 63: 854–868.

PIEROTTI, R., AND C. A. ANNETT. 1986. Reproductive consequences of dietary specialization and switching in an ecological generalist. In press *in* A. C. Kamil, J. R. Krebs, and H. R. Pulliam (eds.). Foraging behavior. Plenum Press, N.Y.

SCHREIBER, E. A., R. W. SCHREIBER, AND J. J. DINSMORE. 1979. Breeding biology of Laughing Gulls in Florida. Part I: Nesting, egg, and incubation parameters. Bird-Banding 50: 304–321.

SOKAL, R. R., AND F. J. ROHLF. 1981. Biometry, 2nd edition. W. H. Freeman, San Francisco, Calif.

TINBERGEN, N. 1960. The Herring Gull's world. Harper and Row, Boston, Mass.

Studies in Avian Biology No. 10:129, 1987.

SEASONAL DISTRIBUTION OF FORAGING GULLS AT FLORIDA LANDFILLS

STEPHEN R. PATTON[1]

ABSTRACT.—The spatial distribution of foraging Laughing Gulls (*Larus atricilla*) and migrant Herring and Ring-billed gulls (*L. argentatus, L. delawarensis*) was studied at two landfills near Tampa Bay during 1981–1982. I determined the distribution of gulls on the active dumping face by dividing the surface into seven regions, observed in the presence and absence of bulldozers. Randomly selected gulls were followed to their landing locations and the species and age-classes of their 10 nearest neighbors were recorded. Log-linear analyses were used to test for random distributions of gulls among species, age-classes, locations, and bulldozer activities. Interactions between species and age, species and bulldozer, species and location, and age and bulldozer were significant for fall, winter, and spring seasons; in summer, only age-related interactions were significant. Consistent positive associations of adult Laughing Gulls with bulldozers, Herring Gulls with the top edge of the face, and Laughing Gulls with the bottom edge of the face were observed; other distributional patterns changed seasonally. Species differences in feeding method, morphology, and behavior contribute to observed distributional patterns.

[1] Department of Biology, University of South Florida, Tampa, Florida 33620.

Studies in Avian Biology No. 10:130, 1987.

PATTERNS OF DISTRIBUTION OF DIURNALLY ROOSTING GULLS IN A COASTAL MARINE ENVIRONMENT

GLEN CHILTON AND SPENCER G. SEALY[1]

ABSTRACT.—Barkley Sound, British Columbia, is an important stopover site for migrating California Gulls (*Larus californicus*) and several other seabird species. The Sound also supports a resident population of the Glaucous-winged Gull (*L. glaucescens*). Most gulls roost communally. The distribution of individuals at roosting sites is a function of prey behavior, opportunities to feed by local enhancement, and species differences in foraging ecology. The average roosting flock of California Gulls, fish-feeding specialists, on 55 transects was 22.5 individuals. As intertidal foragers, Glaucous-winged Gulls occurred on 102 transects in smaller groups that averaged 4.7 individuals. When fish schools apparently became a less reliable resource, flocking tendency decreased in California Gulls, and individuals foraged alone more than they did in flocks.

---

[1] Department of Zoology, University of Manitoba, Winnipeg, Manitoba R3T 2N2, Canada.

Studies in Avian Biology No. 10:131–140, 1987.

HYBRIDIZATION OF GLAUCOUS AND HERRING GULLS IN ICELAND

AGNAR INGOLFSSON[1]

ABSTRACT.—I report here some results from studies during the last two decades on the extensive hybridization between the Herring Gull, *Larus argentatus* and Glaucous Gull, *L. hyperboreus,* which resulted from a massive immigration of *argentatus* to Iceland commencing about 1925. Mate selection was found to be random with respect to primary pigmentation and body size, but site tenacity had an effect. *Argentatus*-like birds raise fewer young per nesting attempt than do *hyperboreus*-like ones, whereas birds of intermediate appearance have a higher incidence of non-breeding than do others. There are also indications that adult *hyperboreus*-like birds survive better than do others. The population is not becoming more *hyperboreus*-like, probably because of continuing low-level immigration of pure *argentatus* from Europe.

Extensive hybridization is taking place in Iceland between Herring Gulls (*Larus argentatus*) and Glaucous Gulls (*L. hyperboreus*). This hybridization is the result of a massive immigration of *L. argentatus* into Iceland which commenced about 1925–1930. The results of field studies of this hybridization up to 1966 are given in Ingolfsson (1970b).

The earlier studies indicated that, following the colonization by *argentatus,* rapid changes in phenotypic traits occurred in gull colonies in southern and eastern Iceland. However, in western Iceland *argentatus* did not make its presence felt to any degree. The aims of the present studies are to follow further developments of the hybridization by investigating selected colonies in western and southeastern Iceland, and to obtain further information on various population phenomena, such as mortality rates and mate selection.

These studies rely largely upon the pattern of pigmentation of the outermost primaries to indicate the degree of "hybridness" of the gulls. In addition, then, an effort was made to study the effectiveness of the "hybrid index," based on primary pattern, by investigating its relationship with other characters.

METHODS

Four gull colonies were studied, all of which had been studied previously. The colony at Bjarnahöfn (or Bjarnarhafnarfjall) is located in western Iceland, while those at Skrudur, Hromundarey and Horn are in southeastern Iceland (see map in Ingolfsson 1970b). At Bjarnarhöfn, samples of gulls were obtained in 1971, 1972, 1978 and 1986 by shooting or by killing the gulls by means of the drug Avertin (tribromoethanol). At other colonies I attempted to capture gulls alive by drugging them with an appropriate dose of Avertin; in this procedure the mortality rate was only about 20%. All adults were tagged with a numbered ring. Those from Hromundarey and Horn as well as a proportion of those caught at Skrudur, were also individually color-ringed. At Skrudur, adults were ringed in 1972, 1974 and 1978; in 1986 a sample of gulls was taken there by killing with Avertin. At Hromundarey gulls were ringed every summer from 1971–1975, and at Horn in the summers of 1971–1973. Additional visits were paid both to the Hromundarey colony for checking of color-ringed gulls in the springs of 1977, 1978 and 1979 and to Horn in 1977. At all of these colonies, chicks were frequently marked as well by use of wing tags and leg rings.

The hybrid index (HI) of all gulls handled was determined. The index, described in Ingolfsson (1970b), is based on the pattern and amount of pigmentation of the five outermost primaries and ranges from 0 (apparently pure *argentatus*) to 5 (apparently pure *hyperboreus*). Values are given to one decimal place. For dead birds the index was assessed in the laboratory and the wings stored for reference. For live birds the index was assessed in the field, and in addition both wings were photographed in color in an outstretched position. Standard body measurements (culmen length from base of feathers on top of bill, bill height at the gonial angle, tarsus length and weight) were taken on all adults handled. The state of primary moult was examined on all dead birds and a primary moult score (PMS), as described in Ingolfsson (1970a), was obtained. The PMS ranges from 0 (all primaries old) to 100 (all primaries new and fully grown). The score progresses rather steadily with time during the moult.

RESULTS

THE RELATIONSHIP OF PRIMARY-BASED HYBRID INDEX TO OTHER CHARACTERS

The effectiveness of the primary pattern as an indicator of hybridness is obviously dependent upon its constancy with age. Photographs were available of 50 gulls captured and photographed in two or more years (interval 1–6 years) in the eastern colonies. Three of these gulls were photographed in three years and so the number of comparisons between successive examinations is 53. The hybrid index had changed in 33 cases. In 28 cases the change was 0.2 HI units or less and the maximum change was 0.5 units (3 cases). There was little indication of correlation between age and HI value. In 20 instances the HI value had lowered with age, while in the remaining 13 cases the index had increased. Changes in pig-

[1] Institute of Biology, University of Iceland, Grensasvegur 12, 108 Reykjavik, Iceland.

TABLE 1

CORRELATION COEFFICIENTS BETWEEN HYBRID INDICES AND BODY MEASUREMENTS OF GULLS AT THREE COLONIES IN EASTERN ICELAND

Locality	Year	Sex	N	Correlation between HI and: Culmen	Bill depth	Tarsus
Horn	1965–1966	♂	16	−0.7099**	−0.8040**	−0.6031**
Horn	1965–1966	♀	19	−0.6052**	−0.2923	−0.7421**
Horn	1971–1973	♂	14	−0.3710	−0.2579	−0.3434
Horn	1971–1973	♀	23	−0.5855**	−0.3584	+0.0108
Hromundarey	1965–1966	♂	41	−0.3619*	−0.0302	−0.2843
Hromundarey	1965–1966	♀	30	−0.1014	−0.3191	−0.1365
Hromundarey	1971–1975	♂	88	+0.0500	−0.2223*	+0.1297
Hromundarey	1971–1975	♀	97	−0.1812	−0.1256	−0.1148
Skrudur	1965	♂	22	−0.1076	−0.1264	−0.1240
Skrudur	1965	♀	19	+0.3137	−0.3836	+0.4614*
Skrudur	1972–1978	♂	180	+0.1254	+0.0589	+0.1314
Skrudur	1972–1978	♀	172	−0.0194	+0.1502	+0.1546*
Skrudur	1986	♂	28	+0.0797	−0.2887	−0.2210
Skrudur	1986	♀	32	+0.1947	−0.1392	+0.3077

* $P < 0.05$.
** $P < 0.01$.

mentation pattern of at least one primary were noted in virtually all cases. Most gulls showed some change in the score of individual feathers, although this did not necessarily lead to a change in the overall hybrid index, since changes in individual primaries sometimes cancelled each other out. The scores for the 9th and 10th primaries tended to become lower with age (their pigmentation tended to decrease), while the reverse was true for the 6th, 7th and 8th; the difference between the two sets of feathers was highly significant ($\chi^2 = 8.28$, $P < 0.005$). Changes in HIs between years, although frequent, were clearly not of such a magnitude as to invalidate the use of the index.

L. argentatus and *L. hyperboreus* differ in body size, the former being smaller, although there is some overlap when birds of the same sex are compared (Ingolfsson 1970b). As body size is likely to be under the control of a greater number of genes than is the primary pattern, an investigation into body size of hybrid populations is of interest. Also, an investigation of the degree of correlation between hybrid index and body size can be expected to indicate the effectiveness of the index as a measure of hybridness.

An overall correlation between body size and primary pattern has been demonstrated using material from a large number of colonies (Ingolfsson 1970b). A closer examination of both old and new data shows, however, that the situation is complex (Table 1). In the Horn colony in 1965 and 1966, where both apparently pure *hyperboreus* and *argentatus* were found nesting, a strong negative correlation between body size and HI was generally found. During 1971–1973 this negative correlation was much less evident, and in some instances significant differences in correlation coefficients between the two periods are found. The HI distribution had at the same time changed markedly (see below). At Hromundarey where no pure *hyperboreus* have been found (probably indicating that hybridization has been taking place for a longer period), the situation is different. Here negative correlation is slight and is significant only for some measurements in males, both during 1965–1966 and 1971–1975. At Skrudur, the largest colony examined and the one where *argentatus* traits are most evident, there is little or no indication of negative correlation between body size and HI. However, there are some instances of significant positive correlations (Table 1). Although this is significant only in the case of females, a similar tendency is seen for the males during 1972–1978. No correlation appeared at Bjarnarhöfn, but the sample is very crowded in the low HI-range.

These results indicate a decrease in correlation between HI and body size as hybridization progresses. This is to be expected if genes for body size and primary pattern are not closely linked. No explanation can be offered for the curious reversal of correlation seen at Skrudur.

Birds from Horn scoring in the low range (HI 0–1.0) are significantly larger than birds scoring in this range from Hromundarey and Skrudur, but significantly smaller than such birds from Bjarnarhöfn. Even 0-scoring birds (i.e., apparently-pure *hyperboreus*) from Horn are small in comparison with Bjarnarhöfn gulls. This may indicate that 0-scoring birds from Horn contain some *argentatus* genes in spite of absence of primary pattern. Interpopulation differences between pure *hyperboreus* populations might provide an alternative although less likely explanation. The Horn *hyperboreus* population would have to contain by far the smallest individuals of any population in the North Atlantic for this explanation to hold (unpublished data). The few birds from Bjarnahöfn scoring above 1.0 are very large compared with eastern birds of similar appearance; some in fact are larger than any bird examined from the east coast.

It is difficult to compare the time of moult of *hyperboreus* and *argentatus*. There are indications of differences in timing among colonies, birds in colonies in southern Iceland moulting a little earlier than more northern ones. There are also differences between years. The gulls examined at Skrudur on June 8–10 1972 were significantly less advanced in primary moult than those examined there on June 8–10 1978 (Mann Whitney U on PMS, $P < 0.01$). In spite of these difficulties the indications are that *hyperboreus* moults its primaries somewhat earlier than does *argentatus* living under similar conditions. Thus at Horn, both in 1965 and in 1966, a significant negative correlation was found between PMS and HI (1965: $r = -0.6518$, df $= 20$, $P < 0.01$; 1966: $r = -0.6480$, df $= 11$, $P < 0.05$) (Fig. 1). No such correlation is found during these years at Hromundarey or at Skrudur. Data on moult for later years are rather scant, except for the Skrudur colony, and give no indication of a correlation between hybrid index and time of moult.

These analyses show that body size and time of moult are genetically influenced. Genes for these attributes are not closely linked with those influencing amount and pattern of pigment on primaries, and a correlation of HI with these features becomes indistinct or indistinguishable after hybridization has been taking place for some time. The HI is at present consequently a poor indicator of the hybridness of individual gulls. It seems, however, reasonable to assume that low-scoring birds have on the average a greater proportion of *hyperboreus* genes than do higher scoring ones. Changes in the HI distribution of a colony with time would thus probably reflect changes in the relative frequency of *argentatus* and *hyperboreus* genes in that colony. A similar argument can be applied to body size, while there are many problems in the use of PMS in this connection.

It is relevant to note that Patten (1980), who studied hybridizing populations of *L. argentatus* and *L. glaucescens* in southern Alaska, found a significant correlation between a hybrid index based on primary pattern (actually a modified form of the index here used) and two other attributes in which the parent populations differed.

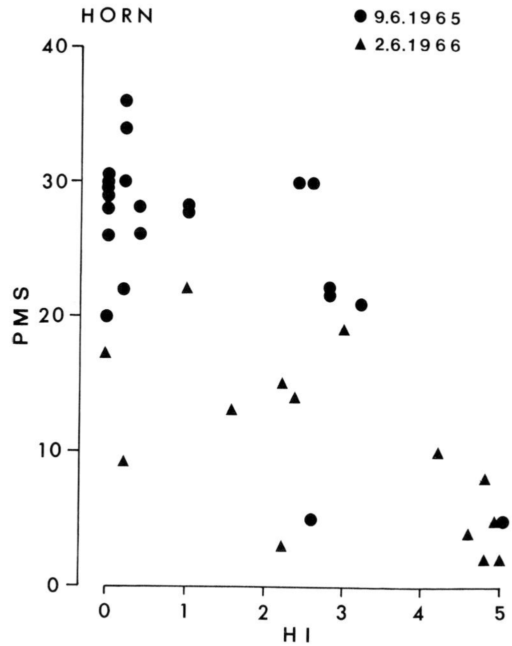

FIGURE 1. Primary moult score (PMS) and hybrid index (HI) of gulls collected on two dates in the Horn colony, southeastern Iceland. A negative correlation is significant for both dates (see text).

The HI Distribution of the Colonies since 1964

Earlier findings indicated a rapid change in average HI of colonies after the immigration of *argentatus* (Ingolfsson 1970b). For the colony at Krisuvík in southwestern Iceland an average increase of at least 0.12 HI units per year is indicated for the 35 years prior to 1965; and a similar value can be obtained for the Vestmannaeyjar colony, also in southwestern Iceland. Furthermore, between 1965 and 1966 a significant increase in mean HI occurred at Horn, and a similar although non-significant tendency was also seen at Hromundarey (excluding the nearby colony of Thvottareyjar from this analysis).

In further analyses of possible changes in the HI distribution birds have been grouped into 6 HI classes. The establishment of these classes has taken into consideration the changes with time in HIs of individual gulls, the overall HI distribution (especially in the eastern colonies), and the form of the relationship between HI and size. The classes are as follows:

class I	HI	0
class II	HI	0.1–1.7
class III	HI	1.8–2.6
class IV	HI	2.7–3.5
class V	HI	3.6–4.4
class VI	HI	4.5–5.0

TABLE 2
THE HI DISTRIBUTION OF GULLS FROM BJARNARHÖFN, WESTERN ICELAND, 1964–1986, SHOWN AS NUMBER AND PERCENTAGE OF GULLS BELONGING TO EACH HI CLASS

HI class	1964–1965 No.	1964–1965 %	1971–1972 No.	1971–1972 %	1978 No.	1978 %	1986 No.	1986 %
I	7	47	35	58	22	61	36	64
II	8	53	24	40	14	39	20	36
III								
IV								
V			1	2				
VI								
Total	15		60		36		56	

TABLE 3
THE HI DISTRIBUTION OF GULLS IN THE A AND B COLONIES AT HORN, SOUTHEASTERN ICELAND, 1965–1973, SHOWN AS THE NUMBER AND PERCENTAGE OF GULLS FALLING IN EACH HI CLASS

HI class	1965, A No.	1965, A %	1966, A No.	1966, A %	1971–1973, A No.	1971–1973, A %	1971–1973, B No.	1971–1973, B %
I	7	32	1	8	1	3		
II	8	36	3	23	3	10		
III	3	14	3	23	4	14		
IV	3	14	1	8	3	10	1	8
V			2	15	6	21	1	8
VI	1	5	3	23	12	41	10	83
Total	22		13		29		12	

On the basis of examination of a large number of museum skins of gulls from outside Iceland, I feel certain that no pure *hyperboreus* will score above 0, while pure *argentatus* may score as low as 4.5, and an occasional aberrant one even lower (see Ingolfsson 1970b for a fuller discussion of this). Results discussed above, however, indicate that classes I and VI may include birds of hybrid origin indistinguishable from pure birds.

Results for the Bjarnarhöfn colony during 1964–1986 are shown in Table 2. At this colony there is no indication of change during the 22 years covered. The non-significant trend is towards lowering of the mean HI with time. Gulls at this colony rarely score above 0.6 and birds scoring 0 remain common.

The results for the Horn colony are shown in Table 3. In 1965 I killed the majority of the gulls, but the colony had regained its size the following year. The HI distribution in the Horn colony was now significantly different (classes I–IV vs. V–VI, Fisher's Exact Test, $P = 0.037$). The replacements were predominantly of classes V–VI, while birds of class I (apparently pure *hyperboreus*) were conspicuously lacking. In 1966 I again killed a large proportion of the gulls. The colony which contained about 20 pairs in 1965 was still of similar size in 1971, but decreased to about 10 pairs in 1973 and had almost disappeared in 1974, after which it has not been investigated. By 1971–1973, the difference from 1965 had become still more marked and the proportion of birds of class VI was now significantly higher than in 1966 (Fisher's Exact Test, $P = 0.014$). By 1971 a few pairs had established a loose colony among Lesser Black-backed Gulls, *L. fuscus*, about 2 km from the old one. The new colony (colony B) contained about 7–10 pairs during 1971–1973. These birds were almost all of class VI. The HI distribution of the two colonies, A and B, during 1971–1973 comes close to being statistically different at the 5% level when proportions of birds scoring above 3.0 are compared (Fisher's Exact Test, $P = 0.057$).

The Hromundarey colony was estimated to contain about 40 pairs in 1965 but in 1971–1979 the colony harbored between 100–150 pairs. Gulls have not been caught here since 1974. No significant changes have been noted at this colony since 1965 (Table 4) and there was no trend apparent during the 4 years 1971–1974.

The results for the Skrudur colony are shown in Table 5. This large colony of about 500 pairs has not shown a clear directional change in HI distribution during the 20-year period 1965–1986. The sample from 1972 is, however, aberrant in that the proportion of class VI birds is low. This sample differs significantly in this respect from the samples of 1974 ($\chi^2 = 5.57$, $P < 0.025$), 1978 ($\chi^2 = 4.56$, $P < 0.05$), and 1986 ($\chi^2 = 11.46$, $P < 0.005$) (recaptures in these years of birds ringed as adults in 1972 are omitted) but not from the smaller sample of 1965 ($\chi^2 = 1.45$, $P > 0.1$). There is no indication of a difference between the sample from 1965 and those from 1974 and 1978. The proportion of class VI birds in the 1986 sample is higher than in any other

TABLE 4
THE HI DISTRIBUTION OF GULLS AT HROMUNDAREY, SOUTHEASTERN ICELAND, 1965–1974, SHOWN AS NUMBER AND PERCENTAGE OF GULLS FALLING IN EACH HI CLASS

HI class	1965 No.	1965 %	1966 No.	1966 %	1971–1974 No.	1971–1974 %
I						
II	9	23	3	10	27	14
III	2	5	3	10	22	12
IV	8	20	4	13	28	15
V	13	33	12	39	53	28
VI	8	20	9	29	57	30
Total	40		31		187	

TABLE 5

THE HI DISTRIBUTION OF GULLS FROM SKRUDUR, EASTERN ICELAND, 1965–1986, SHOWN AS NUMBER AND PERCENTAGE OF GULLS BELONGING TO EACH HI CLASS*

HI class	1965 No.	1965 %	1972 No.	1972 %	1974 No.	1974 %	1978 No.	1978 %	1986 No.	1986 %
I										
II	1	2	2	1	5	5	1	1	1	2
III	2	5	1	1			3	3		
IV			17	11	4 (1)	4	7	6	3	5
V	11	26	50	33	27 (8)	25	27 (3)	23	9	15
VI	27	66	81	54	74 (11)	67	82 (7)	68	47 (1)	78
Total	41		151		110 (20)		120 (10)		60 (1)	

* For 1974, 1978, and 1986 the number of gulls ringed in previous years as adults is shown in parentheses.

sample, but the difference is not statistically significant from any single sample, except that from 1972. In 1978 and 1986, 15 breeders were caught that had been banded as chicks previously at the colony (1971: 6, 1972: 4, 1973: 4, 1974: 1). The HI distribution of these birds does not differ significantly from that of any other Skrudur sample.

Observations in the four colonies show that the situation is now fairly stable with respect to HI distribution. There is no indication that pure *argentatus* is still immigrating on a large scale to preexisting colonies in Iceland. The Horn colony is aberrant in this respect. Birds killed here in 1965 and 1966 were obviously replaced by birds originating elsewhere and the replacements may have included immigrants from abroad. That immigrant Herring Gulls were involved in the formation of the new B colony at Horn is also likely, as the HI distribution there differs significantly from that of known neighboring colonies (except the A colony at Horn) in the high proportion of class VI birds (apparently pure *argentatus*). If this interpretation is correct, immigrant *argentatus* are still arriving in Iceland, but are usually unable to breed unless a vacuum is created by killing gulls in already existing colonies (see Ingolfsson 1978). The B colony at Horn is one of the very few new colonies known with certainty to have been established in Iceland in the last 15 years or so, which further indicates that Iceland as a whole may not have any additional carrying capacity for these gulls.

The observation that the Skrudur and Hromundarey colonies have maintained unchanged but significantly different HI distributions for close to 15 years, although the distance between them is only about 40 km (considerably less than one hour's flying time), shows gene exchange between neighboring colonies to be limited. This is further substantiated by color ringing. Although several hundred adults have been color-ringed in the three colonies in southeastern Iceland, there are only two instances of adults changing colonies, both gulls moving to Hromundarey, one from Skrudur, the other from Horn (55 km away). These are the two colonies closest to the Hromundarey colony.

MATE SELECTION

Previous studies did not indicate assortative matings with respect to HI (Ingolfsson 1970b). New data from the east coast on some 75 additional matings produce the same result: There is no hint of assortative matings with respect to primary pattern.

The data on matings have been analysed with respect to size. Bill height at the gonial angle was used as a measure of size, since sexual dimor-

TABLE 6

BILL HEIGHT AT GONIAL ANGLE IN KNOWN PAIRS FROM VARIOUS LOCALITIES IN SOUTHERN AND EASTERN ICELAND*

Females	Males <20.0 mm	Males 20.0–20.9 mm	Males >20.9 mm	Total
<18.0 mm	3 (4.28)	11 (8.32)	4 (7.39)	18
18.0–18.9 mm	10 (7.50)	20 (18.87)	19 (19.87)	49
>18.9 mm	3 (5.54)	17 (14.32)	16 (16.54)	36
Total	16	48	39	103

$\chi^2 = 5.42$ (df 4), $P > 0.1$.

* Figures in body of table show number of pairs with expected frequencies in parentheses. The expected frequencies were obtained for each colony separately and then totalled.

TABLE 7
THE HI OF FEMALES AND MALES THAT HAVE CHANGED MATES, AND THE HI OF THE MATES LISTED IN CHRONOLOGICAL ORDER

Mate changes of females		Mate changes of males	
Female	Mates	Male	Mates
2.0	4.1	0.2	0.6
	4.7		3.2
2.6	0.2	0.2	5.0
	3.5		2.6
2.8	4.8		0.1
	5.0	1.8	3.2
3.2	3.7		4.7
	2.3	2.3	4.0
3.6	3.9		3.2
	4.1	3.9	3.6
4.3	5.0		3.4
	0.6	4.1	2.0
4.8	2.0		3.6
	4.6	4.1	0.3
			4.6
		4.7	0.8
			2.4

phism is especially great in this dimension (Ingolfsson 1969). All available data have been used except those for the Horn and Hromundarey colonies, where only matings in the year with the maximum number of matings were used. This was 1972 for Horn and 1974 for Hromundarey. The results (Table 6) give no indications of assortative matings. Both Goethe (1961) and Harris and Jones (1969) claim that in *argentatus* the difference in bill size of members of pairs was considerably greater than expected by chance. However, as there was no statistical support for these conclusions, they cannot be properly evaluated. As the data are presented by Goethe, it seems in fact unlikely that statistical treatment would point to assortative matings with respect to size.

Harris (1970) has shown by cross-fostering experiments that imprinting of chicks on parents is of importance in mate selection of *L. fuscus* and *argentatus,* especially as far as females are concerned. I have no direct observations on the role of imprinting in the hybrid population of Iceland, but some information may be obtained by analysing mate changes (sometimes due to my killing a member of a pair). In all I have information on 7 females and 8 males that have changed from one known mate to another (Table 7). There is no indication that primary pattern has anything to do with selection of new mates and it is clear that both males and females frequently have successive mates differing widely in HI. It is therefore unlikely that imprinting on a primary pattern is of significance in mate selection.

Gulls show a high degree of site tenacity, pairs usually returning to the same territory to breed year after year. Gulls changing mates between years show almost the same degree of site tenacity as unchanged pairs. In a high proportion of cases members of newly formed pairs were known to be previous next-door neighbors. Site-tenacity was therefore clearly of importance in pair formation.

NESTING SUCCESS

An attempt was made to estimate the breeding success of birds, with known HI and body size, at Hromundarey. Culmen length was used as a measure of size as it is a less variable indication of body size than weight and was measured more accurately than bill height. The colony was visited frequently during the springs of 1973 and 1974 to check for eggs and chicks. The chicks were tagged as soon after hatching as possible. In 1973 the chicks were followed to a stage very close to fledging (about 4 weeks old), while in 1974 checking was discontinued when most chicks had reached an age of about one week to 10 days.

Some results with respect to HI are given in Table 8. There is little indication of differential breeding success, with one exception: in 1973 a significantly lower proportion of eggs laid by class VI females (apparently pure *argentatus*) resulted in chicks which reached 4 weeks of age than was the case for other females combined ($\chi^2 = 5.31$, $P < 0.025$). A similar tendency is seen among males, although non-significant ($\chi^2 = 2.04$, $P > 0.1$). However, a significantly lower proportion of chicks of class VI males reached the age of four weeks than was the case for chicks of all other class males combined ($\chi^2 = 4.08$, $P < 0.05$). In 1974 the results for females were similar, with close to significantly lower proportion of eggs laid by class VI females resulting in chicks about one week old ($\chi^2 = 3.34$, $P > 0.05$). No such tendency is seen among the males. It should be noted that of the four females of class VI that were investigated in 1974, one also figured in the 1973 sample.

To analyse nesting success in relation to body size, gulls of each sex were grouped into three size classes on the basis of culmen length. The size classes are delimited as follows; an equal number of gulls falls in each class on the average.

	females	males
large	>53.7 mm	>59.7 mm
medium	52.0–53.7 mm	57.5–59.7 mm
small	<52.0 mm	<57.5 mm

TABLE 8
NESTING SUCCESS OF GULLS OF DIFFERENT HI CLASSES AT HROMUNDAREY IN 1973 AND 1974*

		1973				1974			
Sex	HI class	No. nests	Eggs laid	No. chicks tagged	No. 4-week-old chicks	No. nests	Eggs laid	No. chicks tagged	No. 1-week-old chicks
M	II	8	24	14	6	5	15	9	6
	III	4	10	9	4	1	3	2	2
	IV	4	12	8	3	1	3	2	0
	V	4	11	9	4	5	15	9	5
	VI	7	19	15	2	5	14	11	7
	Total	27	76	55	19	17	50	33	20
F	II	4	12	7	4	5	15	12	8
	III	4	9	8	2	3	8	4	6
	IV	5	14	11	5	5	15	11	6
	V	5	13	7	2	5	15	8	4
	VI	6	16	8	0	4	12	8	2
	Total	24	64	41	13	22	65	43	26

* Chicks were usually tagged within 24 hours of hatching.

The results are given in Table 9. In 1973 a significantly smaller proportion of eggs fathered by small males resulted in four-week old young than was the case for eggs fathered by larger males ($\chi^2 = 7.69, P < 0.01$). A tendency for the females in the same direction is non-significant ($\chi^2 = 1.82, P > 0.1$). The proportion of eggs, laid by small females, that resulted in tagged chicks comes closer to being significantly smaller than for larger females ($\chi^2 = 2.44, P > 0.1$). In 1974 the results for males were similar, eggs fathered by small males producing relatively few chicks one week of age ($\chi^2 = 5.61, P < 0.025$). Among females no trend was discernible. It should be noted that of the 4 small males studied in 1974, two were also in the 1973 sample of small males.

In conclusion, the above analyses indicate that the breeding success of gulls with high HIs or small body size is lower than that of other gulls. Both features are *argentatus* attributes. At Hromundarey the negative correlation between body size and HI is slight (Table 1), and the HI distribution of the 20 gulls classified as small appears similar to that of the total sample. Only 6 of the small gulls (30%) belong to HI class VI, the percentage being similar for the total Hromundarey sample (30.5%).

SURVIVAL OF ADULTS AND NON-BREEDING BIRDS

Studies on adult survival and non-breeding birds were done at Hromundarey. The method consisted of a check for the presence of color-ringed birds in years subsequent to ringing. It is felt that scrutiny of the colony was so thorough that virtually all ringed breeding adults were seen. It was, however, clear that all living adults were not necessarily seen in a particular year, since missing gulls would often reappear in later years. During their years of absence these gulls were not breeding, had left after an unsuccessful nesting attempt, or were breeding elsewhere. As re-

TABLE 9
NESTING SUCCESS OF GULLS OF DIFFERENT SIZE CLASSES AT HROMUNDAREY IN 1973 AND 1974

		1973				1974			
Sex	Size class	No. nests	Eggs laid	No. chicks tagged	No. 4-week-old chicks	No. nests	Eggs laid	No. chicks tagged	No. 1-week-old chicks
M	Large	8	22	13	7	7	21	14	9
	Medium	11	27	24	8	6	18	14	10
	Small	7	20	15	0	4	11	6	1
	Total	26	69	52	15	17	50	34	20
F	Large	8	19	18	5	7	21	14	11
	Medium	10	26	16	7	8	24	15	6
	Small	6	18	9	2	7	21	16	9
	Total	24	63	43	14	22	66	45	26

* See text for size-class limits. Chicks were usually tagged within 24 hours of hatching.

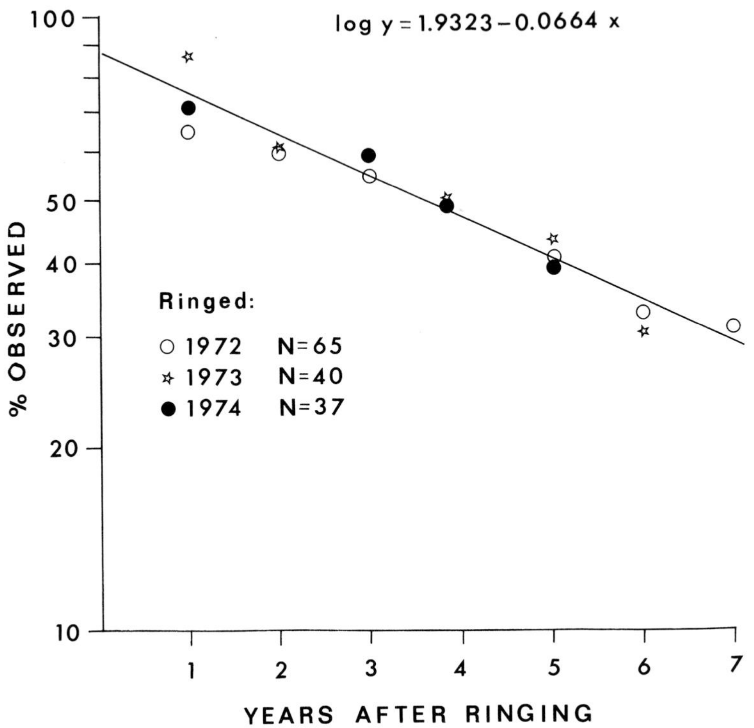

FIGURE 2. Percentage of gulls ringed as adult nesters observed in years subsequent to ringing at the Hromundarey colony, southeastern Iceland. The slope of the line indicates a mortality rate of 14.2% per annum, while the intersection of the line with the percentage axis indicates that 85.6% of surviving gulls are present in the colony in an average year.

corded instances of changing breeding colonies were extremely few and gulls usually relay after a nesting failure, it is probable that non-breeding was the prevalent cause of gull absence. Non-breeding in gulls is recorded frequently by other authors (e.g., Drost et al. 1961).

If it can be assumed that a similar fraction of surviving gulls is absent each year, which seems reasonable, the decline with time of the percentage of ringed gulls observed should be the result of mortality only. The slope of the decline (Fig. 2) indicates a mortality rate of 14.2% per annum (95% confidence limits 12.1–16.2). The intersection of the line with the percentage axis is at 85.6% (antilog 1.9323) (95% confidence limits being 81.7–89.6). This intersection indicates the percentage of surviving gulls present in the colony in an average year.

An adult mortality rate may also be assessed by using all available information to calculate the percentage of gulls known to be alive one year after ringing. This obviously gives a maximum rate, the true rate being possibly somewhat lower because of continuous non-breeding or of movement to other colonies. The maximum mortality rate calculated for 146 gulls ringed in 1972–1975 is 12.3%, a value in reasonable agreement with the values obtained above.

The adult mortality rate estimated at Hromundarey is somewhat higher than that recorded for *argentatus* in several recent studies, in which the rate has been found to be less than 10% per annum (Chabrzyk and Coulson 1976).

When the mortality rate of gulls of various HI classes is compared there is no indication of a differential mortality rate, except where class II gulls are concerned. These *hyperboreus*-like birds (there are no class I birds at Hromundarey) seem to survive better as adults than do other gulls, although the difference does not quite reach significant proportions (Fisher's Exact Test, $P = 0.08$).

There is, however, a difference in the degree of non-breeding (strictly absence from the colony). Table 10 indicates that the incidence of non-breeding is higher in birds of HI classes III–V than in birds closer to *argentatus* or *hyperboreus* in appearance. This tendency is statistically significant for data up to 1978; however, if data for 1979 are incorporated the difference is no longer significant, probably as a result of a smaller sample size (due to mortality).

There is little or no indication of differential adult survival or incidence of non-breeding with respect to body size.

DISCUSSION

The studies reported here from Hromundarey are insufficient to show the relative role of adult survival, non-breeding and nesting success in overall fitness of the gulls. It is possible that the higher degree of non-breeding shown by gulls of HI classes III–V is matched by the lower nesting success of gulls of HI class VI, so the overall selective fitness of these gulls may be similar. But the data indicate that *hyperboreus*-like gulls of HI class II show a selective advantage over other gulls. Furthermore, with respect to size, the data indicate a selective disadvantage of *argentatus*-like birds compared with other gulls. On the average, therefore, fitness of gulls at Hromundarey seems to be inversely related to proportion of *argentatus* traits.

One would therefore expect a gradual change in HI distribution of colonies with time towards a more *hyperboreus*-like condition, but this has not in fact occurred. The situation at the colony at Horn strongly indicates that immigration of *argentatus* from abroad still occurs to a limited degree, based on changing HI distributions. This immigration may be sufficient roughly to balance the selective advantage of *hyperboreus*-like gulls, thus creating a relatively stable situation. The carrying capacity of Iceland for large intertidally feeding gulls seems to have been reached some decades ago; a future invasion by *argentatus* on a scale similar to that of 1925–1930 is therefore not to be expected, unless for some reason the situation in Iceland changes drastically in favor of *argentatus*-like birds. On the other hand, a decrease in the present "immigration pressure" caused by a change in the situation abroad (first and foremost in the British Isles) could be expected to lead to a decrease in the proportion of *argentatus* genes in the gull population of Iceland. It is probable that conditions are different in various parts of Iceland. In particular, future events in the northwest, the stronghold of *hyperboreus,* may well be different from those in the rest of the country. It is planned to continue intermittent observations of the four colonies that have been under study in the past to monitor developments.

TABLE 10
NUMBER OF GULLS IN DIFFERENT HI CLASSES AT HROMUNDAREY EVERY OBSERVATION YEAR FOLLOWING RINGING (A) COMPARED WITH THE NUMBER OF GULLS MISSING IN ONE OR MORE BREEDING SEASONS BUT REAPPEARING LATER (B)

HI class	Data up to 1978		Data up to 1979	
	A	B	A	B
II	9	2	7	3
III	2	1	2	1
IV	7	7	4	7
V	6	11	5	12
VI	16	8	10	11
Total	40	29	28	34

HI classes II + VI vs. III + IV + V: For data up to 1978 (3–5 observation years): $\chi^2 = 5.26$, $P < 0.025$. For data up to 1979 (4–6 observation years): $\chi^2 = 2.34$, $P > 0.1$.

The hybrid situation in Iceland is quite different from that reported by Hoffman et al. (1978) involving *Larus glaucescens* and *L. occidentalis* in the Pacific northwest of America. Although hybridization is extensive and possibly of long standing, mating patterns at the colony studied are assortative, individuals tending to pair with mates similar to themselves. Furthermore, pairs consisting of pure conspecifics hatched significantly fewer chicks than pairs containing at least one hybrid individual. Computer simulations indicate that the situation may remain stable if there is a regular but small influx of pure types.

Still different is the hybridization occurring in southern Alaska between *Larus glaucescens* and *L. argentatus* investigated by Patten (1980). A narrow hybrid zone exists here, with extensive hybridization. Mating patterns are significantly assortative as in the situation noted above including intergrades selecting like types as mating partners, but no difference in breeding success was noted between pure pairs and mixed pairs. It is thought that the situation is maintained both by continuous immigration of parental types into hybridization areas and through changed environmental conditions due to activities of man. Such differences between hybrid situations are of course expected. Each hybrid situation will show its own characteristics depending on a whole range of interacting factors such as age of contact, genetic similarity, selective pressures on populations while allopatric, habitat diversity of contact area, and dispersal patterns.

ACKNOWLEDGMENTS

I wish sincerely to thank the numerous field assistants that have taken part in this study in the last two decades. I would like to thank Richard R. Snell for allowing me to examine gulls collected by him at Skrudur. I have benefitted greatly from constructive criticism of the manuscript by Arnthor Gardarsson, Douglas A. Bell and Wayne Hoffman.

LITERATURE CITED

CHABRZYK, G., AND COULSON, J. C. 1976. Survival and recruitment in the Herring Gull *Larus argentatus.* J. Anim. Ecol. 45:187–203.

DROST, R., FOCKE, E., AND FREYTAG, G. 1961. Entwicklung und Aufbau einer Population der Silbermöwe, *Larus argentatus.* J. Orn. 102:404–429.

GOETHE, F. 1961. Zur Taxonomie der Silbermöwe (*Larus argentatus*) im südlichen deutschen Nordseegebiet. Vogelwarte 21:1–24.

HARRIS, M. P. 1970. Abnormal migration and hybridization of *Larus argentatus* and *L. fuscus* after interspecific fostering experiments. Ibis 112:488–498.

HARRIS, M. P., AND JONES, P. H. 1969. Sexual differences in measurements of Herring and Lesser Black-backed Gulls. Br. Birds 62:129–133.

HOFFMAN, W., WIENS, J. A., AND SCOTT, J. M. 1978. Hybridization between gulls (*Larus glaucescens* and *L. occidentalis*) in the Pacific Northwest. Auk 95: 441–458.

INGOLFSSON, A. 1969. Sexual dimorphism of large gulls (*Larus* spp.). Auk 86:732–737.

INGOLFSSON, A. 1970a. The moult of remiges and rectrices in Great Black-backed Gulls *Larus marinus* and Glaucous Gulls *Larus hyperboreus* in Iceland. Ibis 112:83–92.

INGOLFSSON, A. 1970b. Hybridization of Glaucous Gulls *Larus hyperboreus* and Herring Gulls *L. argentatus* in Iceland. Ibis 112:340–362.

INGOLFSSON, A. 1978. Population phenomena associated with the colonization of Iceland by Herring Gulls. Ibis 120:126.

PATTEN, S. M., JR. 1980. Interbreeding and evolution in the *Larus glaucescens-Larus argentatus* complex on the south coast of Alaska. Ph.D. thesis, The Johns Hopkins University, Baltimore, Md.